The High-Tech Personal Efficiency Program

The High-Tech Personal Efficiency Program

Organizing Your Electronic Resources to Maximize Your Time and Efficiency

Kerry Gleeson

John Wiley & Sons, Inc.
New York • Chichester • Weinheim • Brisbane • Singapore • Toronto

This text is printed on acid-free paper.

Published by John Wiley & Sons, Inc.

Library of Congress Cataloging-in-Publication Data:
Gleeson, Kerry, 1948–
The high-tech personal efficiency program : organizing your electronic resources to maximize your time and efficiency / Kerry Gleeson.
p. cm.
Includes index.
ISBN 0-471-17206-5 (pbk. : alk. paper)
1. Time management. I. Title.
1998

Printed in the United States of America

10 9 8 7 6 5 4 3 2 1

To my wife, Jill, and my children,
Brooke, MacKenzie, and Quinn.

They provide the reason.

Acknowledgments

This book would not have been possible without the support of the worldwide members of the Institute for Business Technology. My thanks to Bary Sherman, Ann Searles, Lena Holmberg, Anders Norberg, Johan Holst, Bruno Savoyat, and Ira Chaleff for providing both constructive ideas and encouragement when I needed it.

I am especially grateful to Jim Robinson, who has worked tirelessly, researching, testing, and contributing to this book. Thanks for being patient with me.

Thanks to my friends in IBT Sweden whose thoughts and experience in helping professionals organize themselves in their electronic environment have been invaluable in the research and writing of this book.

Thanks so much to Carter Crawford and Carol Gorelick for their contribution of Chapter 7 of this book. Chapter 7 was completely researched and written by Carter and Carol. I cannot tell you how much I appreciate both their professionalism and good humor throughout this project.

Jon Ewo, a Norwegian computer consultant, helped considerably bringing clarity to Chapter 6 and other portions of the book. I want to acknowledge his contribution.

Many of our IBT clients contributed their ideas and feedback on the book's content. None were obligated to spend their valuable time making this publication better, but all did so willingly. Special thanks go to Dagfinn Lunde, Duncan Anderson, John Birch, Frank Roland, Philip Dunkley, John Delany, Ruth Maes, Kathryn Ciurczak, and Mike Gallico.

I am grateful to Tom Wrona, whose help with Windows 95 and Macintosh was invaluable.

This book would have been impossible were it not for the wonderful support from my office partner, Tanya Seldomridge. She is the one who had to do both of our jobs so I could spend the necessary time on the book.

My thanks go to editors Jim Childs, Renana Meyers, and Linda Indig; Ralph Butler, illustrator; Rose Design, graphic artists; and Cape Cod Compositors for manuscript preparation.

Contents

Introduction

The purpose of this book is to help you get more done. I am not especially enamored of technology. I am not one of those people who gravitate to the newest gadgets designed to "ease" my life. In fact, "ease" is a word I do not associate with many of these new electronic tools. While I find this to be especially true for desktop computers, it also applies to many other machines we find ourselves struggling with.

In his best-seller, *Being Digital*, Nicholas Negroponte points out, "Many people don't want to use the computer at all. They want to get something done." Isn't it the truth? It is from this perspective that *The High-Tech Personal Efficiency Program* has been written. My experience with helping people get more done with the least amount of effort has taught me there are benefits in embracing these new tools.

I am by nature a lazy person. I like to smell the roses. I do not especially admire people who work long and hard hours. I save my admiration for those who do *not* work long and hard hours but still manage to get done what needs to be done. I've asked myself, "What's wrong with lying back with my feet on the desk, if I have done what I am expected to do?" Nothing that I can think of. *By getting well organized, employing good work practices, and utilizing a computer you will be able produce more than enough to buy yourself plenty of "lazy" time!*

While I'll attempt to cover the use of some of our most practical high-tech tools, *the book's focus will be directed more toward effective work practices.* Changing the way you work and using electronic tools can multiply your personal productivity.

Many of the electrical and electronic gadgets you see today from desktop computers to smart phones are designed to organize you so you can get more done. Excellent idea. But we find that *first it is necessary to organize the tools that were designed to organize you!* What we have learned in teaching others how to use the computer is: Unless the person is well organized in the low-tech environment, the individual is not likely to realize all of the potential benefits available from a computer.

If you have a hard time finding a paper on your desk because your file system is not well thought through, you will have an even harder time finding unorganized electronic documents.

This book is written for nontechnical people who are struggling to keep up with their work while at the same time trying to improve their use of the newer tools available to them. You will soon discover the book's focus is on how to better organize these tools, not on providing you with sophisticated technical wisdom. So this book is not for the techno-wizard. *It is written by a nontechie for nontechies.* In fact, once you have implemented the advice put forward in this book, you will find you have taken only the beginning steps toward becoming competent with technology.

You probably have been encouraged to embrace technology—use a computer, get connected to the Internet, use E-mail. But you know as well as anyone that few of us are very adept at using these technologies. You ask yourself, "How can I get more results, from myself, my staff, and others I work with?" "How can I use these tools effectively?" *"How can I get my people to use their tools more effectively, when I am far from being a techno-wiz?" This book is written for you.* Study upon study shows we have a hard time both understanding and applying technology in our everyday life. From computers to voice mail to programming a digital clock (even with instructions—if you could find them!) life seems to be getting more and more complex. As hard as it is for you to embrace and take advantage of technology in your own work, think of how much more difficult this may be for you as a manager to get other people to use technology effectively.

TECHNOLOGY—THE BAD ALONG WITH THE GOOD!

Interviews with a number of high-level managers in some of the United Kingdom's most prominent companies say their biggest challenge is *information overload.*

Technology speeds up the distribution of information. It also multiplies it. Many of us have been introduced to E-mail. A wonderful invention. Being able to transfer messages via the computer anytime, day or night, has many benefits. It cuts down on interruptions. E-mail protocol permits cryptic messages, typos and all. By connecting to the home server when traveling, you can retrieve your messages and respond to them, often avoiding a pileup of work upon your return. But if for some reason you cannot access your server (try it from abroad on a digital line through the hotel switchboard that's not equipped for new technology) you can find yourself in a mess.

The workplace has changed in many ways as a result of technology, but the most profound way may be in the collaboration required to get things done. The office has become a place to work as a group and not as an individual. *The preferred method of work has become cross-functional teams executing defined projects. Cross-functional work depends on technology.*

Technology itself is often a problem. One manager at **Digital Equipment Corporation** (DEC) returned to the office after a three-week assignment and found one thousand messages awaiting him! "It's some measure of our status how many messages we get, but after a while you just feel swamped," he says.

Anyone can produce an E-mail message and CC (who remembers "carbon copy" anymore anyway?) to tens or hundreds of colleagues about virtually anything with E-mail. In the past there was a physical limit to information sharing, which in its own way sifted the wheat from the chaff. Today, information sharing seems too easy and cheap for too many.

FOCUS ON THE FUNDAMENTALS OF WORK

I recall a time my daughter Quinn (she is 9 years old at this writing) was struggling with homework. She was reading a second-grade science book explaining plants, seeds, stems, and so forth. She would read the section and I would ask her to explain what she'd read. But she said nothing. I mean nothing. Then sadly she muttered, "I am the stupidest person in the world." We went back through the explanations of the terms and found words in the text *I* couldn't define! Out comes the dictionary for second-graders and eventually we were able to understand what she was reading. In school they "taught" Quinn facts, but because they used words she couldn't grasp and never asked if she understood what she was being taught, she didn't get it.

The barriers set up by techno-wizards are enough to squash enthusiasm for the many new tools showing up on the market. Consider computer language. I received **Macmillan Computer Publishing's "New Titles"** issue of the "best choices" in computer books. *"Using GNN* . . . The user-friendly guide to all the basics will help users find the tools they need to get to work immediately with GNN." What's GNN? "Provides information on getting started with GNN and setting up your Internetworks software." What's Internetworks software? *"Using SAP R/3"*; SAP?

Too often, highly intelligent people trying to get a grip on these technological wonders end up feeling like my Quinn did doing that homework—stupid. Try finding a second-grade dictionary of technical terms! You have to wonder. (I should mention that there is a Glossary of technical terms in the back of this book, which I hope you will find helpful.)

The technically inclined may scoff at this book because I am trying to keep as close to the basics as possible. I find most of us have little working knowledge of how computer files are manipulated, how to back up information, or how to design the computer screen to best suit one's method of work. What most of the clients I have met with want to know is what computer tools (software/hardware) are essential? What are the helpful tips that can make one more productive? You will find many of these as a special "Tip" element appearing throughout the text.

The principles you will find in this book are gleaned from the experiences of 150 facilitators who have coached over 300,000 people in some 20 countries on effective work practices. Our job has been not only to teach but to get people to embrace and practice better work behavior.

Okay, we all have our own horror stories about technology, about how to keep up with the amount of information we are expected to process, and about how we're supposed to do this in a "downsized" environment.

So, let's take a look at what you can do about it.

PREVIEW

Getting organized begins by understanding and applying the basics. In this chapter you will learn that:

☞ There are many wonderful high-tech tools on the market. But we suggest you take a critical look at your current (low-tech) work behavior and refine it before you buy high-tech tools. Maybe you won't need so many of the new high-tech toys after all!

☞ While most of us are proficient and technically skilled in our work, too many of us do not understand the principles of work organization or the application of these principles to our job.

☞ Becoming more productive, and therefore more valuable, begins with understanding basic work principles in the low-tech world.

☞ It is unlikely you can ever take full advantage of the benefits of new technology if you do not possess good work practices.

☞ There are key work principles you can tap into that make your life much easier.

1

Principles of Work

I suspect you are reading this book with the hope of discovering ways to better organize yourself with these new high-tech tools you find yourself using. Here is what I have learned about new high-technology tools designed to help our personal productivity:

They almost always turn out to be more complex than you imagined. They require more time and effort to master than most of us are prepared to put into them. They seldom live up to expectations. (Murphy's Laws for the computer age?)

These tools are designed to make us more productive. However, I have seldom, if ever, found that new high-tech tools improved a person's productivity as much as getting the individual to practice good work behaviors based on sound principles in the low-tech environment. Nor have I found that people with poor work behaviors in a low-tech environment are able to take full advantage of newer high-tech tools.

So I will start with the fundamentals of work. Fundamentals have little to do with tools. Fundamentals are about how we approach our work.

- How we plan and execute.
- How consistent and persistent we are.
- How prepared and organized we are.

Some of these fundamentals may already be familiar to you. Some may surprise you. All will make the effort you put into the use of electronic tools that much more rewarding.

ORIGIN OF THE PERSONAL EFFICIENCY PROGRAM (PEP)

In the early 1980s, I was living in Sweden and had a small sales and marketing consulting business. To attract new clients, I devised a compensation plan unique to Sweden at the time: I wouldn't accept a fee unless the client got a measurable result. It had an attractive ring to it, and I found it pretty easy to get companies interested and willing to at least see me and listen to what I had to say. If a potential client thought I had something to offer, the first hurdle to overcome was figuring out what constituted a measurable result. Since I specialized in sales and marketing, I often was able to work out a measurable target, usually increased customers and sales.

The next challenge I faced was creating a marketing and sales campaign that would deliver a measurable result. This was easier than I imagined: All I had to do was *ask* the people who did the work what they would do to produce the desired result. Most of the time *they knew what to do.*

I would then take their input, develop a plan based on it, and give the plan to them. Now this is the interesting part: Almost invariably I would come back and find the plan hadn't been executed. They didn't have time. They had too many other things to do. Someone got sick or went on holiday. This posed a problem for me. I had to get them to do the plan, or I wasn't going to get paid. I found that they were caught up in day-to-day inefficiencies. They were wasting time looking for things, being disorganized in hundreds of ways, and my primary duty became not my expertise in sales and marketing but getting the staff well organized so they could do the things they had been thinking about doing all along.

I succeeded in building up a client base. One of my clients was a branch of Svenska Handelsbanken, one of the most profitable banks in Sweden. They hired me to increase the amount of money in savings accounts. I followed the formula as outlined above and developed the marketing plan. Then the hard part—getting it done.

I found there were several things that kept employees from implementing the plan. For example, as a matter of policy, the bank personnel periodically rotated jobs and workstations. As a result, every few months people found themselves at a new workstation without knowing where things were. It took them a few weeks to get in some semblance of order. Meanwhile, they wasted time.

Instead of processing each transaction immediately, I found some cashiers would create huge backlogs by putting aside until later the tasks they thought would take longer. Cashiers who processed each item of work immediately, as it happened, didn't develop backlogs.

Since there were no baskets on the desks, when mail came in, it sat on the desktop with all the other papers. Sometimes individual items from the day's mail were buried beneath other papers and overlooked entirely.

The manager of the branch was a competent executive, but she spent most of her time dealing with customers. This gave her little time to devote to the question of the organization of the individual workers. I started a standard filing system at each desk and purged the place of clutter. That way, if you had to use a workstation with which you were unfamiliar, you at least knew where to find things. I asked the senior cashier to describe how she processed her work. This became the model for processing transactions in the bank, and the other cashiers began following her model. We set up a central mail center with baskets for each staff member. Soon the workers were initiating their own solutions to common problems affecting their productivity.

Our work in the branch came to the attention of the corporate office and I was asked to package what we had done. It became the Personal Efficiency Program (PEP).

What made PEP unique was its method of implementation. As you can tell from the description above, we focused on the improvement of each individual's work process, looked for areas of waste and for general inefficiencies, and coached the individuals to both introduce more effective work processes and change their own behaviors. We made sure that whatever tools existed were understood and utilized. *But it wasn't the tools themselves that were important. It was how they were used in the context of their work processes and in accordance with reasonable work behavior that was important.*

We have been able to help people become more productive whether they used a paper in box or E-mail. The tools change but the behaviors remain constant.

Over 300,000 people have gone through the PEP program in some 20 different countries since PEP began. And we have learned a great deal from our clients. What we have learned is summarized in this chapter under 13 principles that we refer to as the *Principles of Work.*

WORK PRINCIPLE ONE

Do It Now!

The single most important work principle is **Do It Now!** If your impulse is to postpone what needs to be done at the first opportunity, you will not get much accomplished, no matter what technical tools you use or other work behaviors you may have down pat. The first rule of personal productivity is: *Act on an item the first time you touch it or read it.* This rule, in a sense, flies in the face of conventional time management wisdom. We have all learned to prioritize our work. After all, many of us have learned the hard lesson of working on the unimportant tasks only to have to live with the consequences of not having dealt with that really important project, which blows up in our faces. As bad as that is, there is something worse: the tendency we all have, to one degree or another, to procrastinate—to put off things we do not like to do. The simplest and most effective way to overcome procrastination is to *Do It Now!*

Procrastination is essentially a bad habit. Get into the habit of *Do It Now*, and voilà, no more procrastination! In his book *Getting Things Done*, Edwin Bliss describes it this way:

> When we fail to act as promptly as we should it usually is not because the particular task in question is extremely difficult, but rather because we have formed a habit of procrastinating whenever possible. Procrastination is seldom related to a single item; it is usually an ingrained behavior pattern.

Apply *Do It Now*, and you'll short-circuit the habit of procrastination. *Do It Now* substitutes an action-oriented behavior for the "do it later" behavior. You act before the mental barriers are activated, so you don't have time to think that it's too hard, maybe it will go away, you're not in the mood, maybe someone else will see it, you don't feel like it, and so on.

Most people are very clever, even ingenious, about putting things off. "I don't have time" is a common excuse. "I think they said they're not going to be there today, so I didn't bother to call." "This could take forever, so I had better wait until I have a free day to start." "It's not so important." The list is endless with reasons why a task can't be completed.

My solution is this: *Be as clever about completing things as you've been about putting them off.* So the person is not there. Who else is there who could give you the information? An assistant? Where else could you get this information? To whom could this task be delegated? How can you get this *done*? That is the point, isn't it? How can you get that letter, or that file folder, or that E-mail message responded to so you *never have to look at it again*? That's where you should focus your brainpower—not on clever excuses.

Most importantly, the power behind the concept *Do It Now* comes from the **speed** with which you can get things done. Speed of execution may be one of the few competitive advantages you have control over. Demand a *Do It Now* behavior from your organization and you will see immediate improvement in results.

Some practical tips that may help you get into the *Do It Now* habit include:

If you have work to choose, take the thing you like to do least and do it first. I call it "worst first." It may not turn out to be the highest priority, but you will feel better about getting the most difficult task out of the way and you will be less worried about this "worst task" having to be done at a later time.

If you are not going to do it now, don't read it! If you read it, do it! Don't leaf through the papers in your in tray each time someone delivers mail and don't scroll through your E-mail messages each time the E-mail alarm alerts you. Instead:

Batch routine work. You can get more done! The categorizing and grouping of your work might be called "batching." Each piece of paper, each E-mail message, every telephone call, every interruption that comes your way, and every item you send out is a form of communication. Process similar communications and tasks in batches, reducing waste and motion. You'll complete each task more efficiently.

Work in blocks of time. It is more efficient and effective than working piecemeal. This not only applies to the batching of similar tasks, such as telephone calls or the handling of incoming mail, but also to project work, sales calls, or a marketing campaign. Peter Drucker suggests that the ideal span of time to work is 90 minutes. You will get more done in a concentrated period of 90 minutes than spreading the work over longer periods of time.

Schedule and avoid having to decide. There is an old time management adage: If you want to get something done, schedule it. Schedule times to process batched work. It is important to find a middle point between the extremes of always acting on *everything* as it comes up and habitually putting things off to do later. It is best to schedule different types of work to be done at appropriate times. I refer to this practice as *Do It Now, Later*. Schedule times to do certain work (like your mail) and when the time arrives, *Do It Now*. Don't look at the mail until you are prepared to act. When you do look at it, act on it. *Do It Now, Later*. If you receive a piece of work that would consume time you do not have at the moment, schedule it for an appropriate time in the future (in your calendar) and when the time arrives *Do It Now!*

Build decisiveness into your work habits. Successful people, in general, take little time in making a decision, but take a long time in changing a decision once it has been made. Many people are afraid to be decisive. After all, if you make a decision, you have to live with the consequences. If decisiveness is a weak spot with you, there's an easy way to help you handle the quandary. Simply imagine the worst possible consequence of a decision you are considering, and ask yourself if you can live with that consequence. If the answer is "Yes," go for it. You can't expect to be 100 percent certain of your course of action at all times.

Being decisive has as much to do with deciding what you will not do as with what you should do and when. (See Figure 1.1.) There is no way you are going to get everything done. The volume of information and work we are all being hit with tells us we have to be selective about what we do. I have seen too many people agonize over

Figure 1.1 Do It Now! Being decisive has as much to do with deciding what you will not do as with what you should do and when.

"that stack of reading behind me I never seem to be able to get to." Learn to:

- Say "no" to activities that do not add sufficient value to the end products for which you are responsible.
- Get more clever about utilizing the resources around you to get things done (delegation).
- Filter out marginal information at the source. Get off unnecessary distribution lists and cancel the magazine subscriptions you do not read.

In summary, *make a decision now:*

- Dump it now.
- Delegate it now.
- Designate time or space for it now.
- *Do It Now!*

WORK PRINCIPLE TWO

Organize your materials. Perform the job in an orderly fashion. Clean up afterward.

—4S

Leave it to our Japanese friends to seek out and find useful processes and improve upon them. 4S is the label for:

Sei-li	Organization
Sei-ton	Orderliness
Sei-kez	Neatness
Sei-sou	Cleanliness

4S originated in China. It was successfully copied and implemented by the Japanese, and only recently reached the West. As described by Ingrid Abramovitch in an article entitled "Beyond *Kaizen*" in *Success Magazine* (January/February 1994, p. 85), 4S goes beyond *kaizen*, the better-known quality concept of continuous improvement. *Kaizen* focuses on manufacturing processes, whereas 4S "takes a grass-roots approach, helping each individual to attain the highest level of personal effectiveness."

Concentration on the components of 4S will bring order to your environment and the way you work. With 4S, you address the core elements of improving personal productivity. Organize your low-tech environment, and you will be able to organize the more complex electronic environment.

WORK PRINCIPLE THREE

Continuously improve your personal work process as part of your everyday work.

Whereas the vast majority of people are proficient and technically skilled to do their work, they never learned the principles of work organization or the application of these principles to their jobs.

Over the years, manufacturing firms have spent a great deal of time and money analyzing, refining, and perfecting each step in the manufacturing process. The results show. Productivity and quality have increased dramatically in manufacturing. But in the white-collar work environment, from services to information processing to management, business processes are more difficult to analyze or reengineer. **Personal work processes are rarely even considered to be part of the management business process, let alone analyzed and perfected.**

Because the knowledge of how to process your personal work to achieve both quality and quantity is missing in the white-collar world, we have a missing link in the quality and productivity chain, even in those companies that aggressively tackle these issues.

Most of us have been formally educated to work in our professions, but few of us have been *taught how to work efficiently and effectively.* This is especially true of white-collar workers. Too many white-collar professionals have little idea how to organize themselves or how to best process their work. They may understand how to draw an architectural plan, write a clever ad, or negotiate a deal, but they can't effectively organize their work or cope well with interruptions.

Few of us spend any time improving our work processes and fewer still have any idea as to how to improve those processes, because we haven't been taught any methods for doing so.

On the plus side, experience shows that when the principle of continuous improvement is applied to personal business processes, the results can be outstanding. Results are visible quickly and the fundamentals are easy enough to pick up. Make a practice of continuously improving your work processes and I promise you will be pleasantly surprised at how easy it is and how quickly you will see the results.

WORK PRINCIPLE FOUR

Have a place to put things.

You'd think we all would know and would apply this age-old piece of wisdom, but we don't. I cannot say I was born with the organizational skills I have. I have spent nearly two decades making my livelihood getting others to apply this principle, yet my home might not show it! My twin girls (Kenzie and Quinn) wreaked regular havoc each morning leaving the house for school. "Where are my eyeglasses, Mom?" "Where did you leave them?" "I don't remember." The problem persisted until we identified where to keep the glasses (the top of the dresser in the kitchen) and drummed it home to the kids. This can be more complex than you might think. The location had to be obvious, so it would be easily remembered, yet (because we're talking about glasses) out of harm's way. The kids (and Mom) had to be reminded to put the glasses there until it became a habit. For the first couple of weeks, I would search the house before bedtime to put the eyeglasses where they belonged until everybody began doing it. Compared to identifying appropriate places to put things in the office, this was pretty simple stuff.

In the office, physical restraints, such as furniture, make the situation much more complex. Even the most sophisticated furniture design seems to be wanting. Why do furniture manufacturers insist on having too many small drawers (unsuitable for a filing system) and too few deep drawers where you can set up a system to easily retrieve things? The desktop becomes the place where *everything* is stored, not just the tools needed to produce and what you are working on at the moment. (See Figure 1.2.)

Our Institute for Business Technology Scandinavian offices have participated in the **Digital Equipment Corporation**'s Office of the Future (sometimes referred to as the Flexible Office or Hoteling) concept

Figure 1.2 Have a place to put things *(that you need!)*.

used by a number of large Scandinavian companies. With office space at a premium and the workplace evolving from the office to the road (or the home), companies are retaining us to deal with several issues: First, to design furniture that will be able to store everything needed in a limited space (the concept is to have an open office environment where mobile staff coming into the office building can take their stored work spaces (on wheels) to any open docking station and set up their notebook computers and mobile telephones and get to work). Secondly, to identify what tools and paperwork are critical to keep in this limited space, purging the remainder while organizing what is kept. Thirdly, to introduce proper work practices to maintain good order in the work space. It is possible to design your work space right from the start. Most of us have to work with what we have; that means we must be that much more clever. Inventory your work tools. Identify a place for each.

WORK PRINCIPLE FIVE

Put things back where they belong.

I heard a story about an old-timer who worked for IBM. He would travel to customer sites and repair their mainframe computers. He would often have younger technicians with him on these jobs. He was the butt of jokes because he had "peculiar" clothes and work habits—not your typical IBM "blue suit." He used to wear overalls with dozens of pockets in them. In his pockets were all the different tools he needed or might need. If he saw something to be repaired, even if it hadn't been part of the original repair order, he'd repair it. If he saw a drop of oil on the ground he took out his cloth and cleaned it up—then and there. As he used a tool, he cleaned it and put it back in the appropriate pocket. If a tool broke, he had a requisition form in his pocket that he filled out right at that moment to have the tool replaced. His colleagues would attack the job without bothering to clean up as they went, and inevitably when the end of the day came the old-timer was done and ready to go before the others. This was his way of working. There is much to learn from his example.

One of the less obvious lessons is: **the work is not completed until everything is put back where it belongs *in better shape than when picked up in the first place.*** This is, once again, the 4S process.

Think of it this way: Each piece of work, each task, has a beginning, middle, and end. Part of "beginning" the task must include organizing (planning, preparing for, setting up) for the task. The "middle" is the act of *doing* the task. Finally, along with task completion, the "end" must include putting things back where they belong and improving the condition of everything you touch (including files, tools, etc.).

As you work at a task, whether it's a letter in the office or the salad for dinner, clean up and put back the tools/utensils **as you finish each part of the task!** This works all the better when the tasks are batched. You can plan out and organize all of the tools you will need (whether it's files for correspondence or telephone numbers for batched calls), process the work, and put it all back at one time. Remember the 4S principle to always clean up after the task is performed.

WORK PRINCIPLE SIX

Note things down.

Most people I speak with take a certain degree of pride in their ability to remember "everything" that needs to be done. It is a mental game they play. While this may have been okay at one time, the pace of today's work and home life and the volume of activities we could or should keep up with have grown to such a large extent it has become impractical to expect to keep on top of the thousands of things to be done. No doubt you do remember these "to-do" things, but perhaps not at the time that's most convenient or effective—say, at 3:00 A.M., when you sit up in bed and think, "Oh, I have to take care of . . ." This constant thinking about, planning out, and mental tracking of everything you need to do—this remembering everything you need to follow up on—simply overwhelms most people. (See Figure 1.3.)

Figure 1.3 Note things down.

I don't believe that you necessarily *want* to reinforce your ability to remember the many hundreds of details that make up your workload. Executives and managers should be more interested in *forgetting* about all these things they need to do. Yes, I said *forgetting*. How? **Note it down!**

Preoccupation and Time

Have you ever noticed that the first time you drive someplace it seems to take longer to get there than the second or third time? Have you ever considered why? The first time you drive somewhere you tend to be alert to where you are and where you are going. You are on the lookout for landmarks. "Three blocks past the pharmacy on Hilton Street" forces you to keep an eye out for the pharmacy and count the blocks. Once you have gone someplace a few times, you can drive there and hardly notice the lights go by. You get in the car and the next thing you know you are there! The sense of time has little to do with how fast you are driving. It has much more to do with how alert you are. *When you are preoccupied time flies by.* You have probably experienced starting the workday only to discover it is time for lunch, and you wonder where the morning went and what you got done. Too often the cause of this preoccupation is our attempt to make sense of and keep up with the thousands of things we must do. **It is our attempt to keep on top of all the things we must track and do, mentally.**

I am convinced that this constant, unproductive preoccupation with all the things we have to do is the single largest waste of time and energy, the biggest barrier to individual productivity, and the one thing we can all do something about to materially allow us to take control of our time and work. Begin by noting things down.

- Get a big calendar book or an electronic calendar software package (to write in).
- If inclined to use paper, get a log book (like the old composition books we had in school), so everything you need to remember can be written in one place and become a permanent record you can refer back to.
- Leave your electronic calendar on your computer open so you can enter new activities or changes in status of already recorded activities *as they occur*.

By putting a well-organized system in place, and making a habit of noting down details of things to do, you will have less to remember and far greater ability to *focus* on the job at hand. Focus is the operative word. It is a *major* factor in determining how much you get done.

WORK PRINCIPLE SEVEN

Work tends to fill up the time made available for it.

Parkinson's Law says, "Work expands to fill the time available for its completion." I have seen people spend their whole day only responding to what comes in by way of E-mail and the paper arriving in the in tray. All their time is consumed processing what is there. Ask that person why he/she doesn't initiate, doesn't get around to making those cold calls or put new business on the books, and you will likely hear, "I have no time." While poor organization, procrastination, and working on the wrong priorities can all play a part, there is another dynamic at work.

If you allocate a specific amount of time to complete a certain task you have a much greater chance of getting the work done in that time than if you set no time frame at all. Decide to devote only, say, 30 minutes a day to processing E-mail and paperwork. Schedule one or two times a day to do these tasks (preferably the same time each day). If you set a deadline to complete work you will likely figure out *how* to do it within the time you set for that activity.

What is important to remember is you have as much control over your time as you exert. Leave your time open-ended and you will consume it with low-value activities (we have a hard time doing nothing). Put a time frame around your different work activities and you will be surprised how quickly you begin to get things done.

WORK PRINCIPLE EIGHT

If you have no place to go, any road will get you there.

—NATIVE AMERICAN SAYING

I have always liked this saying. How often have we found ourselves working feverishly only to find it was "getting us nowhere"? A corollary to this saying may be "If you want to get someplace you first must identify a place to go." Once you have an objective in mind, it is then possible to start on the road that will get you there.

Set goals. A goal is a timeless and broad objective or aim, the end to which efforts and actions are directed. *A goal acts as a guiding light.* Goals are potent tools for increased productivity. They keep you focused. With goals comes meaning. Without goals, there is little or no meaning in work and life, and without meaning there can only be dissatisfaction and a general unhappiness.

It isn't enough to have a vague idea of your goals. Goals must be well defined and preferably worked out in writing. To ensure you have a complete set of goals you must first define the different parts of your work. In his book, *The Seven Habits of Highly Effective People*, Stephen Covey refers to the various roles people play. Marketing manager, chairman of the credit committee, XYZ board member, quality circle leader, staff trainer, and so forth can all be considered staff roles you play. Each part of your work life or role you have to play will have its own set of goals. When you see how many goals come into play, it isn't hard to see why people may have difficulty achieving them, particularly when they lack the system required to accomplish their goals.

The beauty of working toward the accomplishment of a goal is that it almost doesn't matter whether you achieve it or not—*the fact that you are working toward things that matter to you is often enough to bring you happiness*. Even the most mundane of actions becomes tolerable, even enjoyable, because you know it's leading you closer to the accomplishment of your goals.

WORK PRINCIPLE NINE

People act when they have a clear picture of what to do.

If you are not acting on something you know you should be acting on, there is a good chance it is because you do not have a clear picture (idea) of what you should do. In his book, *The Management of Time*, James T. McCay writes:

> The pictures in your mind control your actions. If you have no picture; if you can't make out what is going on, you don't act. If your pictures are cloudy and confused, you act hesitantly. If your pictures are clear and accurate, you act definitely and effectively.

Getting clear pictures depends on planning.

WORK PRINCIPLE TEN

Plan it first, finish it fast!

There is an old German saying, *"If you spend fifty percent of your time planning, you will get it done twice as fast."* There is much to be said about this. Let's look at what it takes to make a movie. There are three distinct steps involved in the production of a movie: preproduction, production, and postproduction. Of the three, the most time-consuming element of making the movie is preproduction. The script is only the starting point; the most essential planning document in the preproduction phase is known as the "storyboard." This is a detailed, artistic representation of every single scene that will make up the movie.

Picture a sheet of paper filled only with empty boxes (sometimes you'll even find them in the familiar shape of a television screen). These boxes make up the frames for each scene. Artists sketch in rough outlines to represent what is seen at every point of filming: who and how many people are in a particular scene, what they say, whether a scene is shot in close-up or with a long lens, where the lights are, the step-by-step progression from one shot to the next, and the combination of shots adding up to a single scene; these are all part of the much larger whole—a motion picture.

Why spend so much time and effort on a storyboard? Because one of the most expensive parts of moviemaking is the on-location shooting. Once production is underway, with 200 cast members standing around, you want to waste the least possible time and effort—not to mention money—telling people where they stand and what they do next. That's what *preproduction* is for, not production. With millions of dollars invested, you simply don't waste time when adequate planning and preparation will save you that time and effort.

This wisdom is not new. Decades ago at The Presidents Association, time management guru R. Alec Mackenzie presented a graph comparing planning and execution times. (See Figure 1.4.) The clear advantage of pre-planning is that things get done faster.

If you wish to get more important things done in less time, you must create your weekly "storyboard." Methods of planning you may find useful include:

Daily to-do list. Each day before you go home you should list all of the tasks to do the next day. Your daily to-do list should come from a weekly plan.

Weekly Plan. Once a week you should examine all of your sources of work. By "sources of work" I mean all of your working files, including your projects, your calendar for deadlines, scheduled activities and reminders, your tickler file (call-forward) system for the things that will be showing up during the upcoming week, your pending matters (pending basket, pending files, etc.), any logbook you may keep for one-time tasks, and any other place you may keep or record the things you must do. From these various work sources, make a list of all the tasks to be done in the new week.

Project Plans. Have you ever heard the question, "How do you eat an elephant?" One fellow told me, with lots of ketchup! The answer,

Figure 1.4 Planning and executing two similar groups of projects with different time distributions.

---

though, is one bite at a time, and this is an important tool to get clear pictures of what to do. Take any complex/difficult task or any key objective and break it down to its finest details. How is the task to be accomplished? Who will do it? With what resources? By when? Who is to oversee the project, follow up, and ensure it is done right? Schedule the tasks realistically. Remember, with long-range objectives: *"Little by little does the trick."—Aesop.*

WORK PRINCIPLE ELEVEN

Know what you truly value.

The most important questions you can ask yourself are "What truly matters to me? What principles should I live by?" If you determine what your principles are—those things that you value above all else—your purpose (or as Stephen Covey might say: your mission in life) becomes all the more clear. If you know what is important to you, you can then establish goals to realize it. These goals will be meaningful because achieving them will reward you with what you truly value.

There is tremendous strength in this approach. Charles Hobbs, the author of the book *Time Power*, calls this self-unification:

> When what you do is in congruity with what you believe, and what you believe is the highest of truths, you achieve the most gratifying form of personal productivity and experience the most satisfying form of self-esteem.

By establishing your most valuable priorities in life you can achieve what Hobbs describes as a concentration of power: "The ability to focus on and accomplish your most vital priorities."

By scheduling "what you truly value" in your calendar/diary system you will be working on what is valuable to you, and will be more likely to be satisfied with what you do.

WORK PRINCIPLE TWELVE

Other than a proactive (Do It Now) *habit, follow-through is the single most important ingredient to success in business and life.*

When I say stick to it, I almost literally mean it. Things get done, objectives are made, goals are achieved most often because the person who wanted it stuck to it and made it happen. Calvin Coolidge, the 30th President of the United States, said:

> Nothing in the world can take the place of persistence. Talent will not; nothing is more common than unsuccessful men with talent. Genius will not; unrewarded genius is almost a proverb. Education will not; the world is full of educated derelicts. Persistence and determination alone are omnipotent.

I suspect that your experience tells you this is true. Things happen because you make them happen and/or because you persist until they do. Planning's relationship to persistence can best be summed up in a quote by Napoleon Hill. In his book, *Think and Grow Rich*, he says:

> The majority of men meet with failure because of their lack of persistence in creating new plans to take the place of those which fail.

This is the essence of the work process. Know what you want. Plan how to get it. Act on the plans. Follow up until it happens, or develop new plans to make it happen. Follow up on the new plans over and over until you achieve what you want. How well you do this is determined by how well you are organized.

WORK PRINCIPLE THIRTEEN

Don't do it, delegate it!

What makes highly successful people so successful? I suggest one of the more important characteristics is their ability to get things done through others. Most of us are not skilled at delegating. It is not something we are taught early in our careers. We start out our careers doing the work ourselves. If we are good producers we may eventually end up as a supervisor or manager and we keep doing the work ourselves! It is a difficult transition from worker to delegator. While there are rules of thumb you can use to improve delegation technique, successful delegation begins with a certain mind-set. Instead of approaching work with the question "How can I get this done?" your approach becomes "How can I get others to get this done?"

Because you do not have a direct report does not mean you don't have other resources you can tap into. Suppose you are organizing a New Year's party for the company. Who in the company is a party person who would enjoy being involved? Do you have laundry to pick up? What laundry service delivers to the home? Must cut the lawn? Is there a teenager in the neighborhood wanting extra cash? There are always resources out there. It is up to you to search them out.

SUMMARY

The key to organizing your high-tech tools is to get your low-tech environment in order. Clean up the clutter. Set up simple systems for your low-tech tools (paper files, trays, logbooks, calendars, etc.) that can be replicated in your computer. Make your personal work practices adhere to the sound **Principles of Work.** With the fundamentals in place you will find it much easier to deal with the ever more complex high-tech world. Where to begin is covered in the next chapter.

FOLLOW-UP ACTION POINTS

✔ **Apply the *Do It Now* concept to your work!** If you read or come across a task to be done, Do It Now! If you are not going to process a piece of work, do not look at it. Schedule a time to do it and forget about it. When it pops up on your schedule, *Do It Now, Later.*

✔ **Identify types of work that can be batched** and establish routines to process the work in batches.

✔ Introduce a weekly planning time (best at the end of the week) and **plan out the upcoming week's activities.**

✔ **Write down your goals and objectives**, professional and private. Review these goals as part of your weekly planning process. Add tasks to the weekly plan, so you are consistently making progress towards the accomplishment of these goals and objectives.

✔ **Delegate liberally.** Follow up so delegated items are executed.

PREVIEW

There is a greater likelihood you will succeed in taking control of your high-tech tools if you have control of your low-tech environment. In this chapter you will learn:

- ☞ The best place to begin gaining control over your work processes and tools.
- ☞ How to get your office files in good working order, including what to keep, what to discard, and how to categorize and label what remains.
- ☞ To establish a very few simple routines that will dramatically increase the control you exert over your work.
- ☞ The steps to take to get your office organized like an F-22 fighter plane.

2

Organizing Your Work Space

This book is about getting things done. The simple answer to getting things done is to do them!—to act when you should act. In most cases, the time to act is now. But we tend to act when it is *easy* to do so. And we tend to delay if the task appears difficult. So, if you make it easier to produce, you will produce more. If you are well organized, it is easier to execute and get things done. The better you are organized, the easier it is to work and the more you get done. This applies whether you are digging a ditch or designing an airplane wing.

The computer is a wonderful organizing tool. But if you cannot find it under mounds of paper, it isn't of much use. If you have telephone numbers strewn around the office—in stacks of business cards, on Post-it notes stuck to the computer screen, in your last year's paper calendar system, and taped to the walls of your office—using a computer only creates one more place for you to lose information.

To successfully utilize high-tech tools, you must begin by organizing your low-tech work space.

BEGIN WITH LOW-TECH

Philip is a senior partner in a thriving law firm. He was eager to learn the use of a computer in his work and was even more determined to bring his law firm out of the 1960s into the twenty-first century. The view toward technology by the other senior partners was not nearly as enthusiastic as Philip's. The firm was using an ancient Wang word processing system that, at the time of its installation, was considered state-of-the-art technology.

When I took a walk through the firm I saw that computerization wasn't really the issue. Paper was strewn everywhere. Client files were stacked along the corridors. It won't surprise you to hear that finding client files was one of the biggest complaints. Piles of papers littered nearly every desk of the firm. The papers were mostly old documents from clients, many of which were never responded to. Legal assistants complained that papers sent to a partner for a review sat for weeks and sometimes months without response. Some secretaries didn't have a Wang workstation on their desks and would have to walk to another desk to find a terminal in order to type a document. There was no printer on the first floor (a compatible one for the Wang system was difficult to find), so again the people would have to walk up two flights of stairs to find a workstation to print out documents. Most secretaries and legal assistants couldn't meet with the partners when they needed to, and thus their work would tend to get delayed.

There was general agreement among the staff that they all had too much to do and couldn't possibly be expected to get everything done. The idea of having to find the right type of system for the firm, choose or design software, and then learn how to use this new technology was too much to ask. They could barely keep up with what they had to do already; how could they be expected to take on more, let alone such a daunting project as bringing the firm up to date electronically? In order to move ahead with such a project, the staff had to see how much better their work life would be with modern systems to support them. They also had to be given resources now, in the form of time, energy, and help, to get caught up with existing work so they could give adequate attention to learning and converting to any new system brought in. It was highly unlikely any new system was going to be even found and designed, let alone implemented, in such an environment. Here is what we did.

CLEAN OUT THE CLUTTER

We began by forming small groups of staff (partners, secretaries, legal assistants, etc.), with approximately six people to a group. The group members were given a brief orientation regarding what we hoped to accomplish (get staff better organized so they had more time and less work to do) and how we intended to go about it (get the clutter cleaned up, put tools in good working order, and set up systems to make it easy to find things). They were given instruction booklets and sent to their work places to get started. First step, clean up the clutter.

What is clutter? Clutter is the mess you face every day when you walk into your office. It's your coat that's flung over the back of your office guest chair because you didn't hang it on the coat tree this morning. It's the half dozen reports perched on the corner of your filing cabinet and buried under the remains of yesterday's in-office lunch. It's the stack of magazines you haven't gotten around to reading. It's the mound of outgoing and incoming mail strewn across your desk. It's the unfinished letters you are writing by hand to give them a personal touch. It's the cassette tapes you meant to take home to listen to over the weekend that are now buried under the quarterly budget. It is all the reminders spread over the desk and walls about all the things you must do.

I should note here that there is a school of thought that claims the best way to work is by having all projects visible for constant reminders. It even has its own name: "information persistence." Information persistence assumes that constant reminders of what you have to do stimulate creativity. One of the largest furniture manufacturers in the world has taken this concept and designed office furniture with extensions for placing each ongoing project such that it is in your face all day. I cannot agree that this is particularly healthy. For some it will be true that working in that environment may be conducive to creativity. But most of us begin to not even notice what is around us after a while—especially if what we are being reminded of becomes overwhelming!

This is how we tackled it.

Once back in the office, the workers began by going through *every single bit and piece of paper on their desks or anywhere near their working spaces.* First they would start by taking all of the supplies and papers out of their drawers and piling them on the desks. This included contents of file drawers, baskets where paper existed, stacks of unfiled business cards, coffee mugs from bygone marketing campaigns, outdated business letterheads, annual reports from the 1980s, and so forth. You name it, they were encouraged to pile it on their desks for critical review.

Why so dramatic a start? It has been said a famous explorer burned his ships upon arrival in the New World so his crew would be that much more motivated and have that much more courage to succeed in their mission. The process I describe above is analogous in that unless you force the issue and give little choice, most of us will not push through this rather boring task with the speed and enthusiasm needed to get it done quickly. Pile all of this on the desk, preferably on top of current important work, and watch how fast you can process it!

Each person was instructed to pick up the first piece of paper, identify it, and determine what must be done *to process it to completion*. They could:

1. Deal with it until completed.
2. Deal with it as completely as possible and then place it in a "pending" basket if very short-term, or a "tickler" file system (a system of 43 hanging folders labeled 1–31—a number for each day—and 1–12 for each month) under the appropriate date, if awaiting a response or scheduling the work for a specific future date.
3. Delegate it.
4. If it's needed for ongoing work, put it in a pile on the floor of papers to file in the "working files" defined later in this chapter.
5. If it's information that has value but that requires no action at the moment, put it in another pile on the floor of papers to file in a "reference files" system (defined later in this chapter).
6. If it has historical or legal significance, but is seldom referred to, put it in a pile on the floor of papers to file in an "archive files" system (again defined later in this chapter).
7. Create a pile on the floor of any papers that belong in the client files.
8. Throw it away! Do this if it's trivial, of no use, already dealt with, or exists elsewhere. This was often one of the hardest tasks for the participants. Some would argue, "I am **the** person in the firm everyone comes to to find things. If I throw

something away, I am sure to be asked for it." Good point, but it should be noted that the office was strewn with extra copies of nearly all documents, so it could have been retrieved elsewhere. If you ask yourself the question, "Is it possible I am going to need this?," the likely answer is yes, because it is almost always possible. I suggested they instead follow Stephanie Winston's advice from her book *Getting Organized* and ask themselves the question, "If I needed it, where could I get it?" If you can figure out where to get it, why keep a copy? The end result was that upwards of 50 percent of the papers in the offices went into the trash.

Because people often overlook the obvious, all participants were instructed to be alert to the fundamentals a white-collar worker deals with every day including desk, pens, tape, paper clips, staples, lights, chair, computer, filing systems, binders, computer disks, and so on. It's not uncommon to walk into a person's office and find these items in disarray—scissors misplaced, staplers broken, tape dispenser empty, and papers randomly scattered. Yet somehow we expect to perform our work effectively in this condition. All we asked was that they step back and take a good look at their office environment. Was the desk set up most suitably? Was the office warm enough in winter and cool enough in summer? Was the chair comfortable? The cleanup process resulted in a complete inventory of what was in the office (paper, tools, supplies, furniture, etc.) as well as its operational state. This inventory dictated the resources we had to work with and the restraints we had to work within.

SET UP SYSTEMS AND CONTROL POINTS

Once we had a good idea of what we would keep, we could then set up an effective operating system for what remained. This paper-based system became the model used for the organization of all of the offices' tools, including the computers that eventually were to be purchased. With the paper-sorting process done, we were able to compare what remained in stacks with the resources the office staff had to work with. Most people had at least one deep drawer in their desk for a file system. Some didn't so we had to improvise. If they had shelf space we set up the system in binders. Everyone was instructed to organize their papers and files by frequency of use. Those things used most frequently needed to be near at hand. The desk is a work surface, and the only papers on it should be those being worked with now.

Figure 2.1 gives an overview of how we set up the individuals'

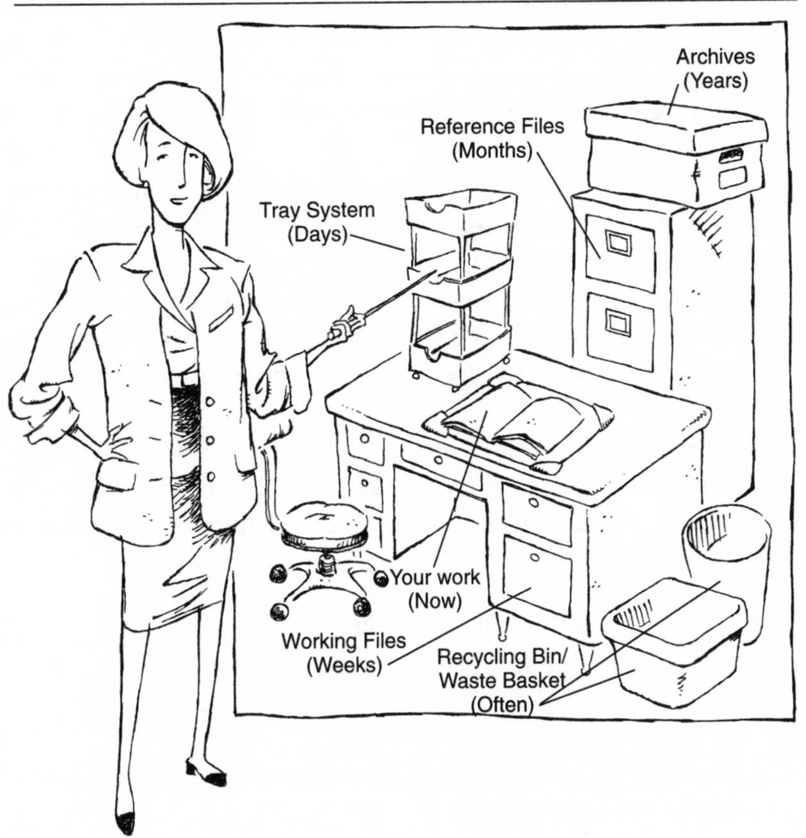

Figure 2.1 Paper control points.

work stations. We began the process of organizing the work space using this paper control points illustration as a guideline.

Get three trays or baskets and mark them "In," "Pending," and "Out." The in basket will receive all *new* material. The pending basket is for those things *you cannot do now*, for things *that are out of your control, usually awaiting someone else's input.* The out basket is for all those papers that have been *completed but not yet delivered.* It should be noted the out basket is not a place to put papers to be filed; it is best to file them immediately. Some find it useful to have a fourth tray for reading materials. The in, pending, out, and reading baskets (trays) are for tasks completed over the course of a few days at most. (See Figure 2.2.)

Set up three separate file systems: working files, reference files, and archive files. These three files are vital *paper control points* for managing your work flow.

It is vital to develop a file structure that embraces all your work and is easy to conceptualize. When this is done, it is easy to decide where to file a document and where to look for it when you need to retrieve it—whether it is a working, reference, or archive file. You do this by mapping out your key responsibilities and the activities and information required to accomplish these tasks. (See Figure 2.3.)

Organize other media. Other items that also need to be well organized are books, shelves, briefcase, address book, and business cards, to name a few. Participants adapted the principles of organizing to these other media by:

1. Grouping similar things together.
2. Placing them in their own space or container.
3. Labeling them clearly.

Since legal cases often accumulate huge amounts of paper that won't fit easily in the deep drawers of a desk, it was necessary to devise a different system to store active cases in a person's office. Large wire mesh baskets were placed on available flat surfaces (not

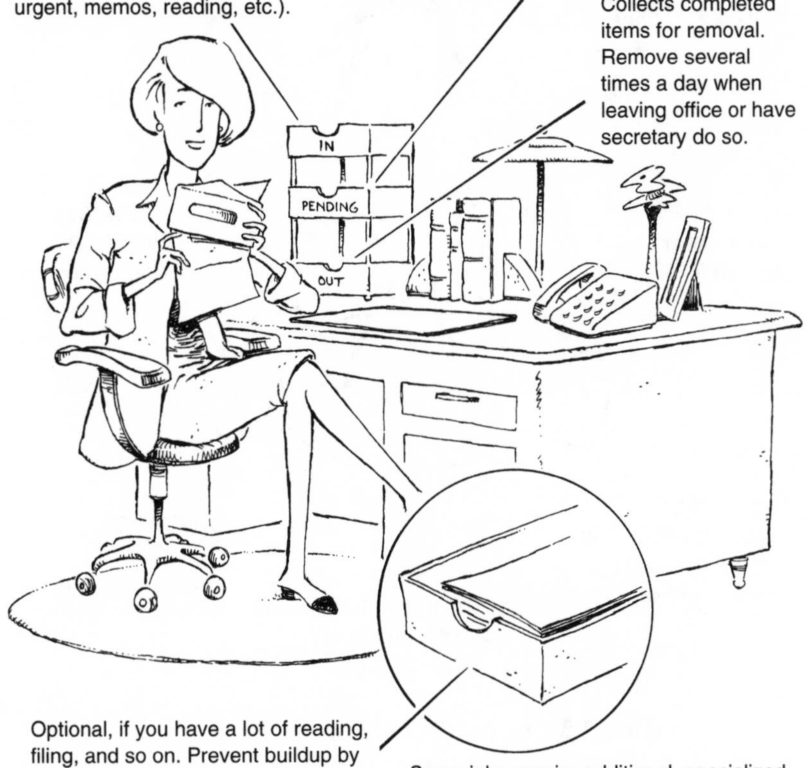

Incoming mail and notes, never before touched. When you pick something up, act on it! If you have an assistant, mail should be screened and sorted into folders that delineate your priorities when you are rushed (e.g., signature, urgent, memos, reading, etc.).
In, Pending, and Out trays must be within arm's reach for efficiency.
Short-term pending, for things you have tried to act on and couldn't complete (e.g., awaiting info, awaiting call-back, interrupted for more urgent matter). NOT FOR: PROCRASTINATION, INCOMPLETE PROJECTS, OR TICKLER-FILE ITEMS.
Collects completed items for removal. Remove several times a day when leaving office or have secretary do so.
IN
PENDING
OUT
Optional, if you have a lot of reading, filing, and so on. Prevent buildup by reading short items at once, scanning table of contents and clipping articles, sharing reading load across department, and clipping or summarizing, scheduling a time for regular reading.
Some jobs require additional, specialized trays to facilitate the work flow.

Figure 2.2 The tray system.

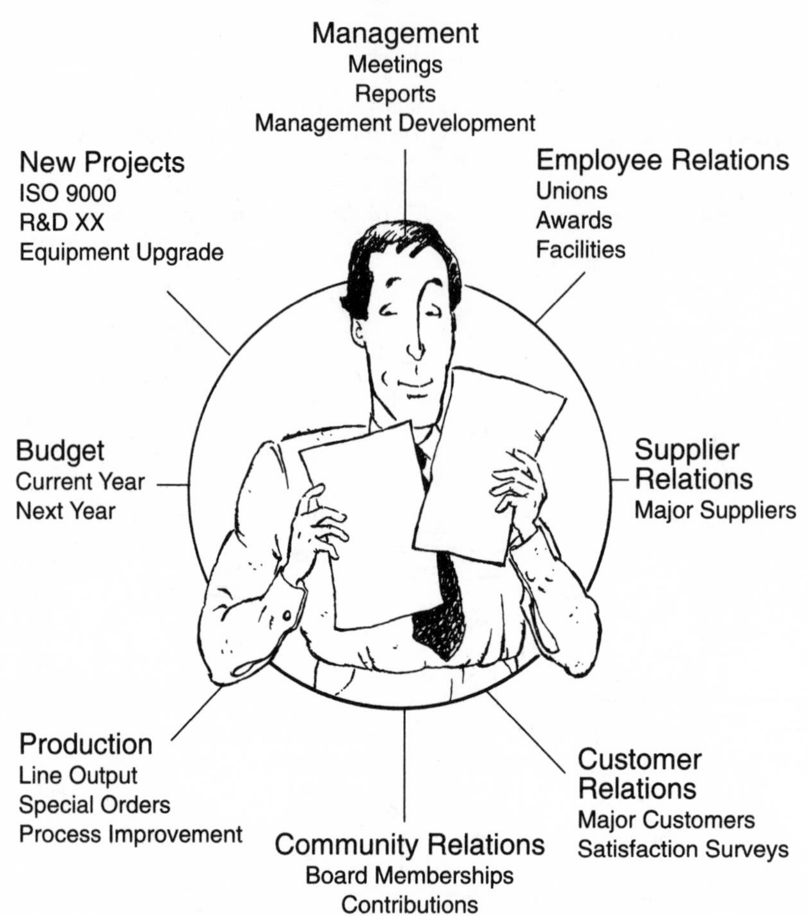

Figure 2.3 Responsibilities map for files structure.

the person's work desk), and all papers related to a case were placed in a basket and labeled. Each case had its own basket.

Supplies drawers were organized by using plastic trays in the drawers to hold and separate the supplies. Any missing supplies were noted and purchased. Name and address lists were updated. Those partners using a computer were encouraged to set up their address book in it. However, the majority of the staff simply updated their paper address books (in some cases they had to go to an office supply store and get one). Bookshelves were put in order. Depending on the number of books, some went as far as categorizing the books and organizing the shelves accordingly.

Briefcases were cleaned out and organized based upon use. Loose software application diskettes were gathered together along with the accompanying documentation and placed in the original software boxes and stored in a section of their bookshelves for easy access. (See Figure 2.4.)

Working, Reference, and Archive File System Details

I will now go into more detail on the setup of the working, reference, and archive file systems.

Working files usually contain six types of information (see Figure 2.5):

1. *Fingertip information.* These files contain phone lists, address lists, computer codes, company policies, and so on—things you refer to frequently and want at your fingertips when you need them.

2. *Items to be discussed.* There should be a file for routine meetings and a file for each staff member with whom you interact.

3. *Routine functions.* These files contain information that you need for routine tasks performed daily, weekly, or monthly. You might, for instance, have a file called "Trips" in which you would store papers, airline tickets, passport, and so on needed for an upcoming trip. You might include a file called "Meetings," a place to store papers needed for one or more upcoming meetings.

4. *Current projects.* These are the projects you're working on now. Create a hanging file for each project and include anything necessary for your current work.

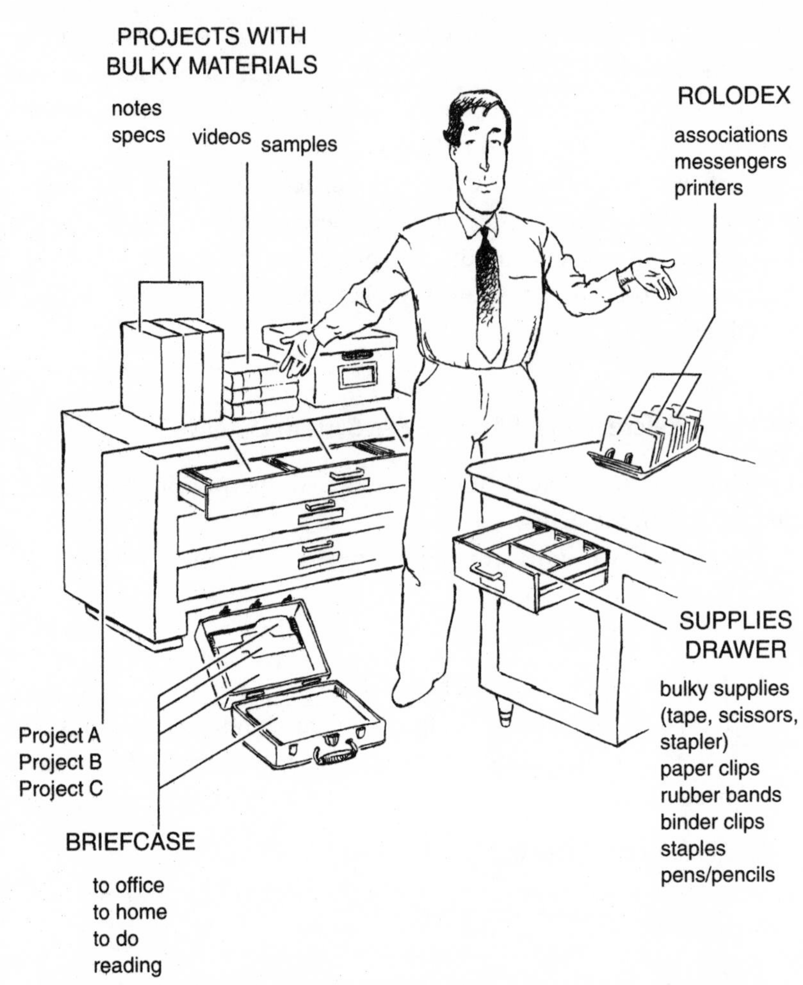

Figure 2.4 Adapting the principles of organizing to other media.

5. *A tickler file* (call forwarding system). This file is usually divided into two parts: One set of folders is numbered 1–12, representing the months of the year; the other set is numbered 1–31, for the days of the month. The tickler file is used for longer-term pending and follow-through items. Items requiring

Eighty percent of your work involves twenty percent of your files. These are your active files that you will be using over the next few weeks. Keep these separate from other files and within arm's reach. This is the most important part of your paper control system.
WORKING FILE
ISO 9000
R&D XX
Upgrade
Projects
Budget
Customers
Community Relations
Employee Relations
Management
Production
Suppliers
Fingertip Information
Tickler
Batch Files
Routine Functions
Meetings

Use your responsibilities map (Figure 2.3) to create your working files structure. There may be several file folders behind each hanging file heading that relate to the broad category.

Figure 2.5 Working files.

action will be filed behind the appropriate 1–31 day section in the current month. The file is checked each day for items requiring action that day. Items requiring action after the current month would be filed in the appropriate month folder when they will require action. At the end of the current month, you would remove items from the next month's folder and file them behind the appropriate dates in the 1–31 section.

6. *An alphabetized file system (A–Z)* for the remaining ongoing work. To avoid a big pending file, set up the alphabetized file system that makes odd tasks easier to find, both for yourself and for others. Because you can find things under the appropriate letter, you avoid needing to constantly create new hanging files, labels, and so forth.

Reference files contain these items:

- Research for your future projects.
- Past projects to which you refer.
- Resource information.
- Personnel information.
- Administrative data.
- Budget information.
- Account records.

When setting up reference files (see Figure 2.6), consider these two things:

- What information you want to retain.
- How you can best organize your reference files for ease of retrieval.

The following ideas might help you in determining how to structure your reference files:

Figure 2.6 Reference files.

1. Create a list of subjects (topics) from the papers you have decided need filing. Compare this list of subjects with the key components of your job (e.g., contracts, trade fairs, product development, budget, personnel, etc.). Group these subjects into broad categories. These become the categories in your reference files.

2. Label file folders clearly and appropriately, based on the categories you've identified.
3. Cull existing files and throw away useless paper.
4. Use hanging files to organize drawers with one or more categories.
5. Alphabetize files within categories or subcategories.
6. Label file drawers and file folders with large, clear letters that make retrieval and refiling easy and fast.

Archive files are most often set up for common use, so how they are structured and categorized may be different from how you might choose to set up your personal files. There may, in fact, be two systems required: an individual one in your office for archival documents strictly related to your job, and a second system outside your office for common use. You need to ask yourself some questions to determine how to set up the best archive system.

- Do you have departmental archives? What about company archives?
- What is the policy on document retention?
- Who is responsible for maintaining archives?
- Does an indexing system exist?
- What are the procedures for retrieving documents from archives?
- Can you rely on documents being recovered if needed?
- Have you tested this system lately?

I find that no matter how complete a departmental archive system might be, there's almost always a need for some form of personal archive system in the office. The personal archives can reside in the cabinets furthest away from the desk, since they will be the least referenced files.

Filing: Categorizing and Labeling

The main purpose in filing *anything* is to be able to find it again. The easiest way to do this is to create broad, general categories that will be genuinely useful and easily understood by others. A good rule of thumb is to set up file systems not only so *you* can find things, but also so *anyone else* can find them. The reason is twofold: (1) on occasion somebody else may need to find items in your files, and (2) if it's simple enough for someone else to use, it's probably simple enough for you! You will probably need to subcategorize within a broad category, but the main idea is to have general categories. Label your drawers as well as the files within them, using lettering that is big, bold, and easy to read. In naming files, use the word(s) that first come to mind. This is what typically will trigger retrieval.

Tips for Filing

The following suggestions will make your filing system more efficient (see Figure 2.7):

- Use hanging files. Hanging files support files better and facilitate refiling in the correct place. Box-bottom hanging files can hold several manila folders on the same subject.
- Label files with large, clear letters. This facilitates retrieval and refiling.
- Align label tabs according to categories and subcategories. This allows the eye to more efficiently scan to the correct file. Categories can be given colored labels to further aid scanning. (Make sure this step is necessary and useful. I was told of one secretary who spent an entire day color-coding the supervisor's files only to discover the supervisor was color-blind.)
- Create an index for large reference files. This enables a manager, for instance, to retrieve files easily in the secretary's

Figure 2.7 Filing tips.

absence. It also minimizes duplicate files and coordinates the use of shared files.

SUMMARY

When we were done purging each office of unnecessary papers and getting what was left into tip-top shape, the difference was startling. Partners and staff alike appreciated the time and effort it took to get their offices organized.

The new system was visible. As a result, we were then able to move into the electronic environment and replicate the paper-and-desk organizing method on the computer. It is much easier to create a system for tangible objects (paper, desk, books, etc.) than for the intangible representations of advanced electronic tools. Create your system in the low-tech environment and use that as your model for your electronic environment. The principles apply in both cases. In the next chapter we will cover how to transfer this system of organization to the computer.

FOLLOW-UP ACTION POINTS

✔ Your task is to **clear the backlog and organize your work area.** Most likely you will require a day or more to do this. Schedule this so you can work on it undisturbed, if possible.

✔ If you have not done so already, **get four trays and mark them "In," "Pending," "Out," and "Reading."** Your in basket will receive all new material. Your pending basket is for those things you cannot do now, for things that are out of your control. Your out basket is for all those papers you've completed. Your reading basket is for all reading matter you wish to take care of during a set reading time.

✔ **Empty out onto your desk every piece of paper or document from your drawers, trays, walls, and briefcase.** Look everywhere—under the blotter, behind the curtain, under the desk. Take no prisoners!

✔ **Pick up the top piece of paper and deal with it now.** You should do one of the following:

a. Deal with it until completed.

b. Deal with it as completely as you can and then place it in the pending basket if very short-term or the tickler file under the appropriate date while awaiting a response.

c. Delegate it.

d. File it in your working files, if it's needed for ongoing work or projects.

e. File it in the reference files, if it's information you need but which requires you to do nothing at the moment.

f. File it in the archive files.

g. Throw it away! Do this if it's trivial, of no use, or already dealt with. Use the aforementioned information retention criteria to help you make this call.

✔ Now, **make a list of all the things you've had bouncing around inside your brain to do**—the small bits and pieces of bad conscience, the not-so-urgent items, the little things you may have been putting off. Dump it all out of your head and onto paper. Once you've listed everything, start acting on each item, beginning at the top of the list. Don't stop until you've worked your way through each and every item on your list. That's how you'll know when you're done.

✔ **Set up your working files, reference files, and archive files.** These control points are vital to controlling the flow of paper in your office and, therefore, vital to your work flow.

✔ **Create a tickler file** if haven't already done so. If you have a secretary, the tickler file should be maintained at the secretary's desk.

✔ **Make a list of missing supplies and tools necessary for you to do your job**—tape, staples, extra file folders, labels, pens, formatted disks, scissors, envelopes, stamps, and so on. Make sure you have them all on hand, and that everything works.

PREVIEW

The computer needs to be organized like any other tool with storage capacity. In this chapter you will learn:

- ☞ How to replicate your office organization on the computer.
- ☞ How to categorize and organize your computer document files.
- ☞ How to organize your computer desktop for easy access to your computer's applications and document files.
- ☞ How to organize other documents such as E-mail and electronic faxes for easy retrieval.

3
Organizing the Computer

"***You mean I have to organize the tool that is supposed to organize me!"*** Now you say, "I am organized! I have a desk and I can see the desktop. I have the most important information in a hanging file drawer organized so it is easy to access. Everything has its place." You have succeeded in the world of atoms. Good start! But you are not yet complete. You still have to deal with the ever more complex digital world.

In this chapter we will begin by organizing the tools designed to organize you (Figure 3.1). Simply put: We will mirror the organization of your desk on the computer!

There's more. Working in the digital world requires different systems, routines, and behaviors than we are used to. You need to know how best to integrate information and work with the setup on your computer.

With the organizational basics in place from Chapters 1 and 2, organizing the computer should be a piece of cake! Read on.

Figure 3.1 Organizing your computer.

THE PROCESS

By organizing the computer we mean:

- Creating a file system for the document files in the computer.
- Creating a file system for retained E-mail and fax messages.
- Having these systems mirror the organization of the rest of your information.
- Transferring those documents you wish to retain into their appropriate electronic folders.
- Creating a computer desktop (the screen that normally shows up when first turning the computer on), that makes access to files and applications easy.

By systems and behaviors we mean finding:

- Effective ways to cope with electronic overload.
- Effective ways to find electronic information.
- Ways to protect your electronic files.

The process itself is fairly straightforward, but it does not appear very natural for people to extend their organization of files and work to the world of the computer. Maybe it is because paper is more "visible" than electronic documents.

CASE IN POINT

Even when clients have completed the Personal Efficiency Program process at their desks, they do not necessarily apply the principles to their computers. The same law office client described in Chapter 2 made little progress applying the PEP principles in the electronic environment. Many partners purchased computers and several began using Lotus Organizer (a personal information manager [PIM] program used as a calendar and to track tasks). The firm also brought in outside help to train staff on how best to use Lotus Organizer. Yet, when I looked into the state of their electronic organization, none of the steps they had taken at their desks were applied to their computers.

John, a partner at the firm, had put a great deal of effort into converting to a digital-based work system from a paper-based work system. He successfully incorporated the principles of PEP into his office and his working relationship with his assistant. He purchased a state-of-the-art computer system and got up and running with Windows 95. But when it came to organizing the computer—nada.

His view was that organizing his computer wasn't going to speed up the amount of information he needed to process his work. He could always add more and more files to his desktop. It wasn't as if he had a storage problem; today's computers seem to have unlimited storage capacity.

This seems reasonable enough until you see the study by Account Temps, International, reported in *The Wall Street Journal* a year or so ago, of professionals' work habits and the difficulties they experience finding things in the office. The study reported professionals spend, on average, six weeks a year looking for things! I'll bet that sounds a bit exaggerated. But nevertheless, I think it's true. Your most compelling reason to spend the time and exert the effort to set up a structured file system is precisely because electronic files are intangible and storage is practically limitless. Without the basics in place, you'll find yourself electronically overwhelmed!

In the law office, John wasn't especially interested in the technical side of the computer. He was interested in getting things done. Many of us would like to avoid having to learn about the inner workings of a computer. After all, one can say it isn't essential for you to understand the mechanics of an engine to drive a car. (Only professional drivers might argue they wouldn't be able to do what they do if they didn't understand the inner workings of the tool of their trade.)

This attitude is prevalent, even in early schooling. My children are learning math in school with a calculator. According to my eldest, Brooke, "It's easier!" If you can divide and multiply with the press of a button, is it necessary to be able to understand and work through the mathematical processes to solve problems without a calculator? A European friend of mine thinks so. He feels so strongly about it, his kids do their math on paper. "How can they really understand it if they cannot work through the problem themselves?"

Understanding the components and the basics of how a computer works will translate into increased productivity. This is not to say that understanding the mechanics of a computer is more important than practical usage of the computer. Our objective is to provide enough grounding on how the computer works so you will be able to put it to better use.

COMPUTERS 101

It's important to have a clear understanding of the basic components of a computer and some of the common computer terms we will be using in this chapter if you want to take full advantage of the computer as an organizing tool. (See Figure 3.2.) If you are unfamiliar with a computer, I recommend you go to the Glossary at the back of the book and look at the terms listed there. Skip this section if you are confident about your basic computer terminology.

Computer Features—What to Consider When Purchasing a Computer

I assume you have a computer. If you do not or you intend to upgrade to a newer computer, there are some rules of thumb to guide you through the process. In general, you should begin by determining what you want to accomplish with the computer. If the main purpose for the computer is to do word processing or spreadsheet activities, a less sophisticated machine would be needed compared to, say, a need for multimedia features. When purchasing a computer you want to consider:

- Hard drive, the computer's storage space for applications and files, needs to have sufficient space to hold all you will store in the computer. If you store graphics (Power Point Images, etc.),

Figure 3.2 Know the language!

or download big files from the Internet, you will need a hard drive of several gigabytes.

- Random-access memory (RAM), the memory used by the computer to run applications and perform computer tasks, needs to be sufficient to speedily process what you tell the computer to process. Graphics programs hog memory. Nearly all current applications require much more memory than in the recent past. If purchasing a desktop or notebook computer, purchase a minimum of 64 megabytes of RAM.
- Computer processor also determines how quickly the computer processes your commands. At this book's publication date, I wouldn't consider purchasing a computer with less than 233 MHz Pentium (or equivalent brand).
- Computer monitor/screen—if purchasing a notebook with a multimedia purpose in mind, you will want an active matrix

screen for the best possible color clarity. Notebook screens of 12″ to 14″ are now available.

- Modem, that device which allows connection through the telephone to other computers, should be of sufficient baud rate to speedily transfer information. This is especially critical if you download large files from the Internet. The higher the baud rate, the better. If you travel outside of the country, consider ClipperVom World PC Card Modem, which automatically configures itself to be compatible in both your home country and the countries you may go to with your computer.
- CD-ROM, that device which allows the computer to read from a compact disc, is quickly becoming a must-have device for a computer. Most software applications come only in CD-ROM format nowadays. The speed with which the CD-ROM works becomes an important consideration.

Network Computers

For many individuals, I am sorry to say, it will be impossible to organize their files and documents as suggested in this chapter. If your computer is on a network with a central machine (server), it will likely have common files structured and organized by the technology department based upon its own protocol. Even if you planned on setting up personal files, the computer on your desk may or may not have its own hard drive. Your company may discourage the use of personal files on the network.

The structure I am suggesting applies well to individual PCs. But the working, reference, and archive system does not work well in a network file structure.

Experience teaches that most companies who have network computers and servers allow for both a structure for common files and a Drive: devoted to personal files. If that is the case with your company, then it will be possible for you to set up your personal files according to the structure advised in this chapter and use the company structure for common files available to all staff.

CREATING A FILE SYSTEM FOR DOCUMENT FILES IN THE COMPUTER

Where to Begin—The Computer's Operating System

It is through the operating system on the computer (be it DOS—an early version of Windows—or the current version, Windows 95; OS/2; Mac; or another operating system) that application and document files can be found (Figure 3.3). Each operating system has its own set of commands or icons that allow you to manipulate files and organize them for easy use. The first step toward organizing John's

Figure 3.3 Know your operating system!

computer information was learning more about his operating system's file management protocol.

File management in most current operating systems isn't too difficult. DOS can be the most complex. But it's a Windows world out there, and chances are most of you are using Windows (as John was). Begin by getting a working understanding of your operating system's file management. Run through the tutorial. Search out file management in the systems manual and help section and study up. If you are like me—lost, slow to learn, and possessing modest skills in only one language (English—nontechnical)—you might seek out a computer coach (someone technically skilled with a working knowledge of your operating system) to coach you through the learning process. I find it highly worthwhile, even though coaching can be costly. In fact, this is the role I played with John. I was able to get John familiar with his operating system's (in his case, Windows 95) features and commands as we organized his files. With or without the benefit of a coach, your first action is gaining familiarity with your machine's operating system. Before you start messing with your computer files, know what you are doing or find someone who does.

Step One: Back It Up!

No matter how skilled you may be with file management, it is wise to do a complete backup (make a copy) of your hard drive before you begin deleting and reorganizing your files (Figure 3.4). Each operating system has a way to back up files, some easier than others. In Windows 95, it is as simple as going into Microsoft Tools in the Start menu, selecting Backup, and following the instructions. Have plenty of formatted blank disks or tapes available, and as you use them, label them. See page 91 for efficient backup systems.

Step Two: Naming Electronic Files

The design of your computer file system should mirror your paper file system.

A typical computer user might have the following software applications, the documents of which will require organizing:

Figure 3.4 Back up often!

1. A word processing application and the documents created and saved.
2. A spreadsheet application and the created spreadsheets.
3. Internet browser software, saved E-mail messages, and documents downloaded from the World Wide Web.
4. A fax application and saved fax messages.
5. A calendaring application with to-do tasks and reminders.
6. Possibly a groupware application and its databases.
7. An application for personal and business finances.
8. A project planning application and the project plans.

9. A shared database and its files.
10. A graphics application with saved images, overheads, and so forth.

Each of these applications produces electronic documents that, in all likelihood, have matching categories in your paper-based systems. If you took Chapter 2 to heart, you will have purged and organized your papers in working, reference, and archive file systems. We find it is best to begin computer organization by mapping out the structure of your paper files. An electronic equivalent to Figure 2.1 is a good place to start. See Figure 3.5.

Begin by making a list of the categories of your paper-based system. Use the form in Figure 3.6 for direction. Go through your working files and reference files and list the categories. Save this list for use later.

Figure 3.7 may make this more clear. The first tier, Mail file folder/directory structure, represents the broad categories 1working,

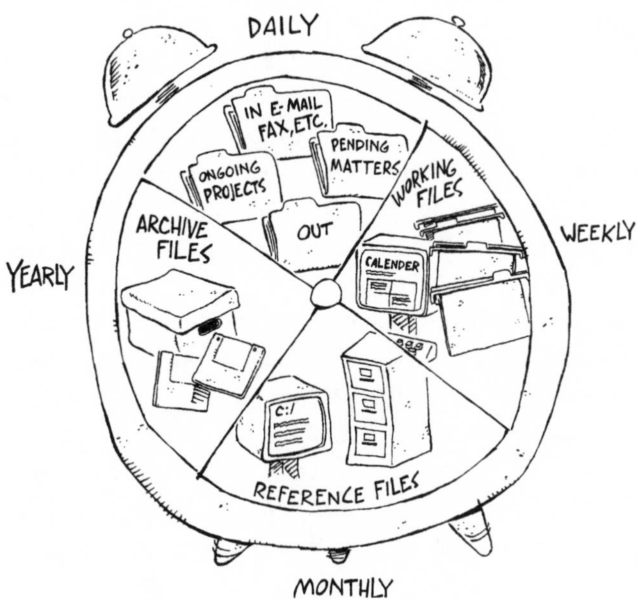

Figure 3.5 Electronic control chart.

Sample Work Sheet

Major Responsibility *(from map)*:

Category	Subcategory or Files	Files		

Figure 3.6 Do your own thing.

2reference, 3archive. The typical documents found in each broad category are outlined in the next two tiers. Use this to better define where you might place your electronic documents.

Keep in mind that the intention is **not** to relocate the program application files, each of which resides in its own folder/directory (WPwin60, Excel, Approach, etc.). The idea is to relocate the files that you, the user, create with these programs (document.doc, xlreport.xls, database.dbf, etc.) so that you can maintain and retrieve these documents more easily. It is very important that you move *only* those files that you create and avoid moving program files. Moving, changing, or losing program files will adversely affect the operation of the program, and may necessitate reinstalling the program.

If the file that you are attempting to open has an identifying extension (.doc, .zip, .xxx), then whenever you double-click on that type of file, Win 95 will automatically open it using the program that you indicated. So the smart/efficient thing to do might be to use the dot three extensions to identify the type file as produced by the software. The added convenience

1st Tier	2nd Tier	3rd Tier	4th Tier
Main file folder/directory structure These folders/ directories are set up using numbers as the first character to ensure their placement at the top of the hard disk tree structure.	**Responsibilities** This folder/ directory tier should be general headings resulting from the responsibility map and will not contain specific files.	**Specific names subdir and files** The names chosen for subdirectories must be general in nature with each tier giving more information about the files in the grouping. When the number of files in a group becomes excessive, start thinking about creating further subdirectories of the existing group.	**Files** Careful thought is necessary when choosing a file name. Use only abbreviations that are meaningful to you and instantly recognizable. Be consistent with your format and name files so that you will have some idea what it is if you see it away from its home folder/ folder/directory.
Working Files *(folder or folder/ folder/directory)* **1working**	**Clients** **Customers** **Finances** **Forms** **People (Personnel)** **Pending** **Projects** **(etc.)**	Actual client/ customer names Expense record, budgets Form names or numbers Actual names of people or personnel records Items awaiting completion Projects currently in progress	
Reference Files *(folder or directory)* **2reference** or **2refernz** (Win 3.x/DOS)	**Graphs / Charts** **Spreadsheets** **Completed projects** **Expense reports** **Reports** **Evaluations** **(etc.)**		
Archive Files *(folder or directory)* **3archive**	**Previous year's tax returns** **"Must saves"**		

Figure 3.7 Organizing the hard drive with Windows Explorer (Win 95).

is the ability to open a file by simply double-clicking on the file's icon. Windows 95 will then open the appropriate program to run that file and then open the file also.

Step Three: Organizing the Computer's Hard Drive

Now that you have gained a working understanding of your operating system's file management, made a backup of your hard drive, and drawn up a list of your working and reference paper system files, you can begin to organize your computer files. As I mentioned previously, each operating system file manager has its own set of commands to maneuver the files on the hard drive. I will show how to arrange your files according to the PEP principles for Windows 95 and will indicate the differences in three of the other most widely used operating systems—Windows 3.1, DOS, and Apple Mac.

Here is how the law firm's computer files were organized. It should be noted that the law firm had no local area network (LAN); each computer was stand-alone, and therefore the organization of the files was done on the PC's C: drive.

PLEASE NOTE: If your computer is part of a LAN, there may be a policy or advisory against storing data on the local (or C:) drive. If your documents must remain on the LAN server, do not attempt to reorganize those files without first clearing it with your technologies department.

Organizing the Hard Drive with the Windows 95 Operating System

To organize the hard drive with Win 95:

1. Create three folders (directories) on the C: drive. Name them **1working, 2reference, 3archive.** Working, reference, and archive electronic file categories allow you to store your document files in a system identical to your paper-based files. Putting the numeral "1" next to your most important file category (working) will position it at the top of the C: drive tree. The numeral "2" puts the reference files below the working ones and the number "3" puts archive last. In Windows Explorer, **highlight Ms-dos_62 or Hard Disk C:**

(or the C:). Choose **File, New, Folder. Type** the name **1working** and press **enter.** Repeat the process and create the 2reference and 3archive directories.

2. Create a shortcut from the 1working folder to the 2reference folder and a shortcut from 1working folder to the 3archive folder. (See Figure 3.8.) One of the nicest features of Windows 95 are shortcuts. Shortcuts are most often used on the desktop for quickly accessing files and applications. They work like this: Software applications must be able to locate the document files you create. Normally applications are limited to seeing document files in one directory only, but if the application is directed to look for its document files in the 1working directory, and you've created a shortcut to the 2reference directory and the 3archive directory, the application will be able to locate its document files in all three folders at the same time.

   To create a shortcut from your 1working folder to the 2reference folder return to **Windows Explorer** and locate and **highlight** the **1working folder. Click on File** and **choose New and Shortcut.** In the **dialog box type C:\2reference. Click** on **Next** and **type in Reference Files** for the name of the shortcut. Follow the same procedure to create a shortcut from the 1working to the 3archive folder, naming the shortcut Archive Files.

3. Using your prepared list of paper files, create a matching set of subfolders in the 1working, 2reference, and 3archive electronic folders. As a Windows 95 user, you will know 95 has an improved file name capability over earlier Windows versions and DOS. Windows 95 allows more than 250 characters in a file name. DOS and earlier Windows versions limited the file name to eight characters plus a three-character extension. This made file naming difficult at best. Windows 95 makes it possible to create an identical set of files in your electronic working file directory as you have in your paper working file system.

   To create subfolders in the 1working, 2reference, and 3archive folders return to the **Explorer**. (See Figure 3.9.)

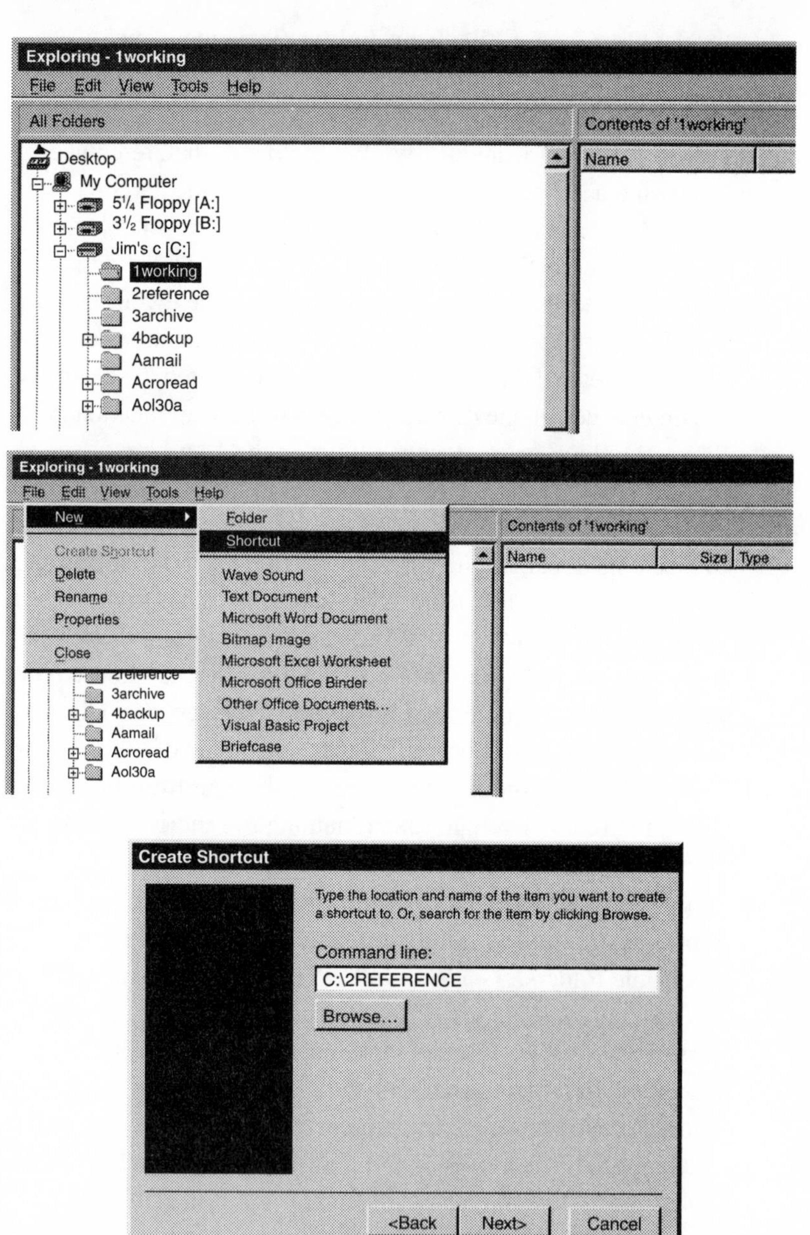

Figure 3.8 Creating shortcuts from 1working folder to the 2reference and 3archive folders.

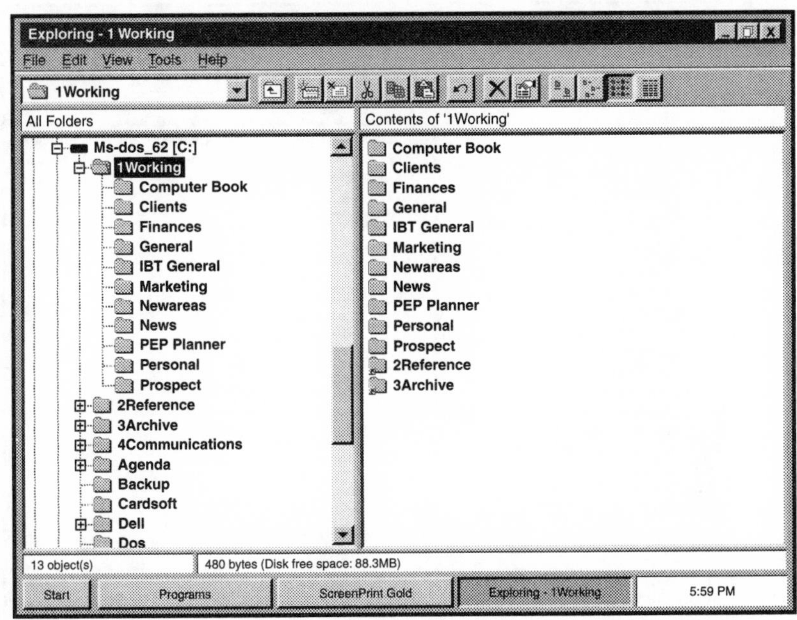

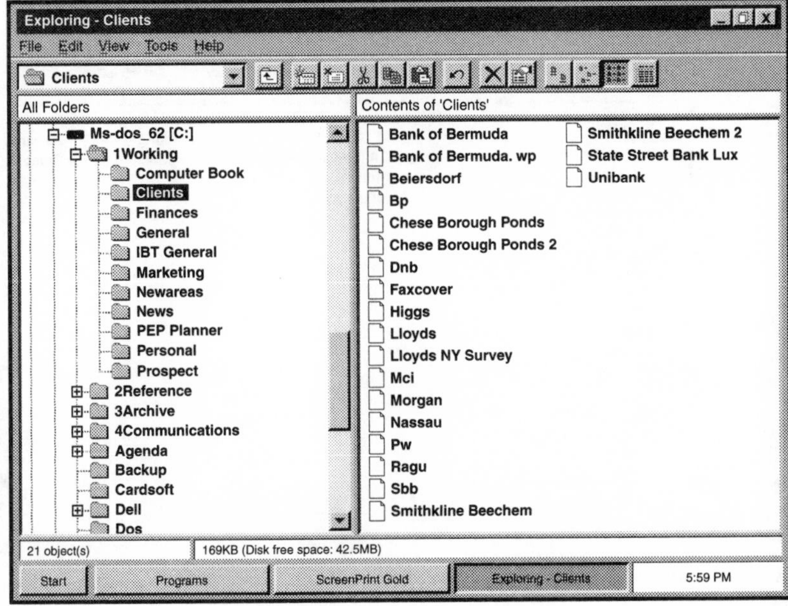

Figure 3.9 Creating subfolders.

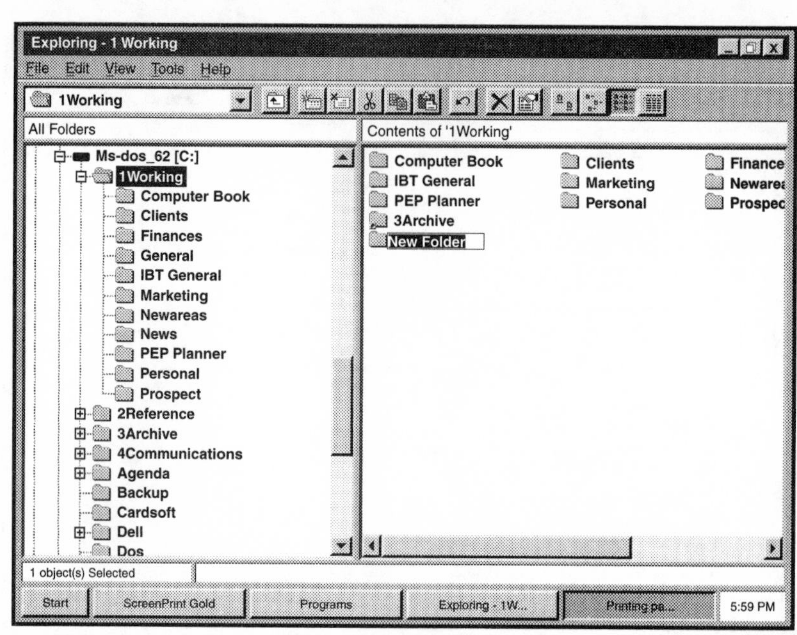

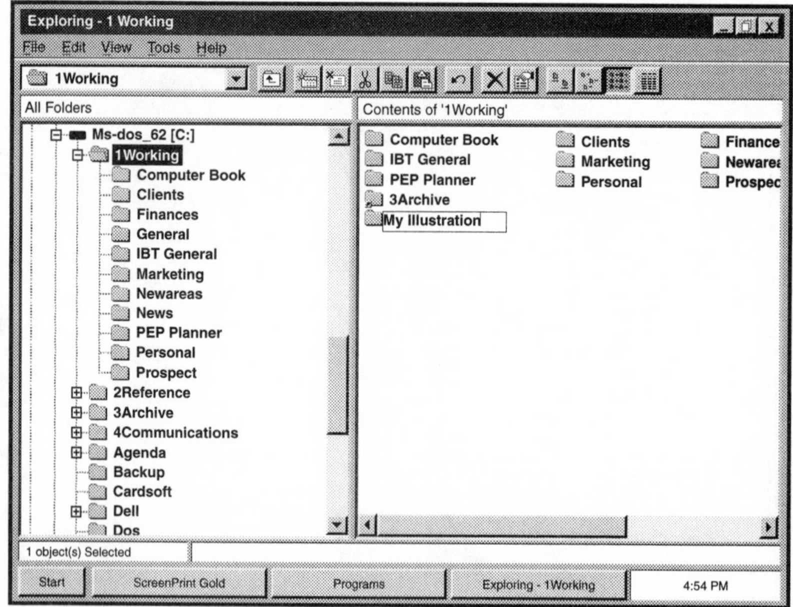

Figure 3.9 Continued.

Highlight the 1working folder. Choose **File, New,** and **Folder. Type in the first name from your paper Working files list. Press enter.** Repeat this process for all categories on the working files list. When the working files list is complete, move on to the reference files list and follow the same procedure. **Highlight 2reference,** choose **File, New, Folder,** and **type in the category name and enter.** Do the same for 3archive.

There is one problem to consider. If you are using non-Windows 95 applications (earlier 3.1 Windows applications or DOS applications), only the first six characters plus a tilde (~) and a number will show up in the applications directory. If, for example, you create a file in Windows 95 for your non-Windows 95 WordPerfect application called Marketing Documents, it will show up as Market~1 in the WordPerfect directory. In naming a file in non-Windows 95 applications, keep the old naming conventions in mind. You might name the file Marketng.doc. Better yet, upgrade your applications to take advantage of Windows 95. Figures 3.10 and 3.11 show how documents look in the Windows 95 Explorer and the WordPerfect for DOS directories.

4. Change the software application default settings so documents are placed within the 1working, 2reference, and 3archive directories. Most of the recent applications (word processing, spreadsheet, graphics, etc.) allow for users to designate the path (address) where the created documents will be stored. Many applications have set the default to search for created documents in the applications directory. Assuming you have created a shortcut to a Windows application on your desktop, the easiest way to redirect applications in Windows 95 is to go the **Win 95 desktop, right click on the shortcut that launches the application, select Properties, click the Shortcut tab, go to the Start In field and type C:\1working and save and exit Properties.** (See Figure 3.12.) You can always look through the applications configuration options for redirecting the default settings of the document files.

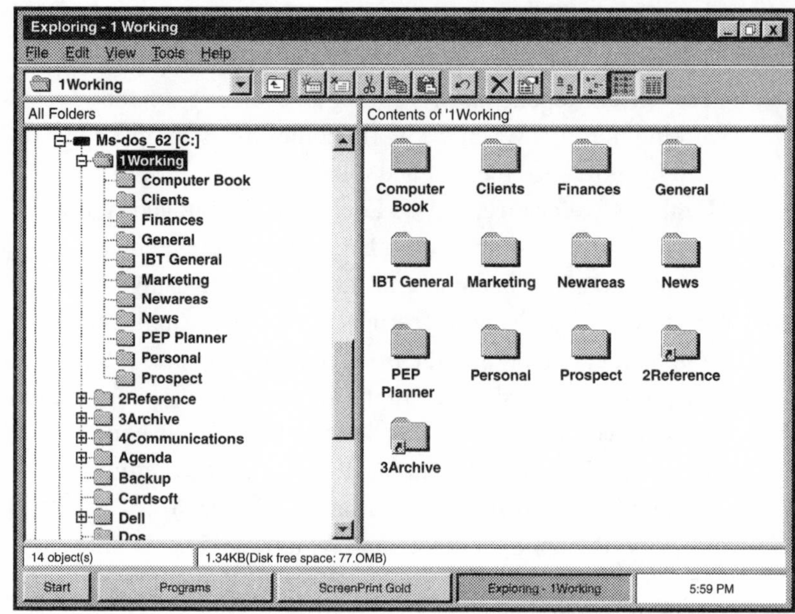

Figure 3.10 Documents in Windows Explorer (Win 95).

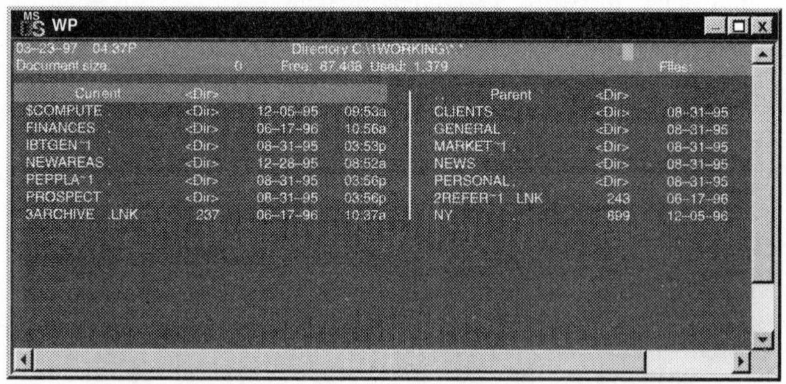

Figure 3.11 Documents in WordPerfect for DOS directory.

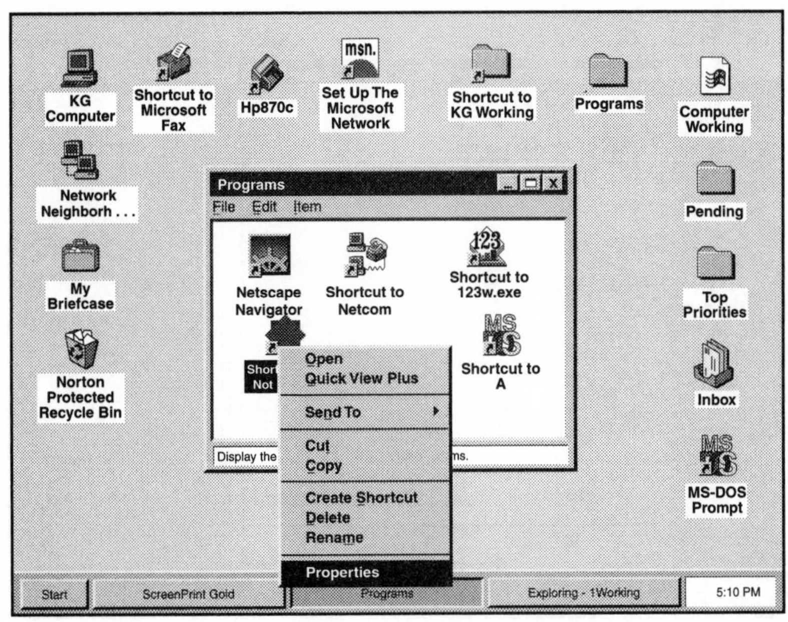

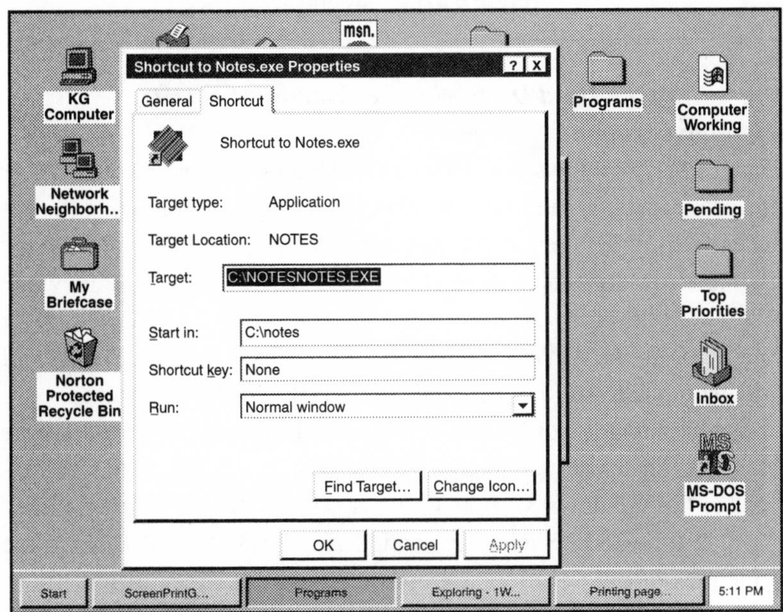

Figure 3.12 Redirecting applications.

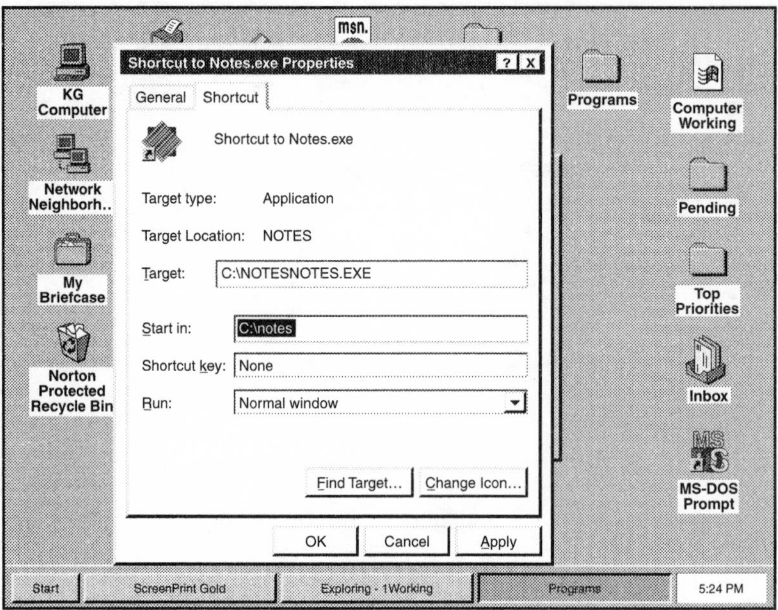

Figure 3.12 Continued.

Organizing the Hard Drive with the Windows 3.x, DOS, and Mac Operating Systems

If you have Windows 3.x, DOS, or a Mac operating system, follow the procedures as outlined previously, noting the following differences: Win 95 has long file names while Win 3.x and DOS restrict the file names to eight plus three characters. Win 95 has shortcuts while Win 3.x follows the scenic route and DOS goes the long way over rough terrain. However, despite the differences, the PEP principles of organization are still possible and beneficial with these systems.

The counterpart for Win 95's Windows Explorer in Win 3.1 is File Manager. The closest counterpart in DOS is DOS Shell. In the illustrations of the Windows 3.1 File Manager organization (Figures 3.13 and 3.14) and the DOS Shell organization (Figure 3.15), one can see the principles of organization at work. Although the advantage of shortcuts is not available, the organization and close proximity of the files makes browsing to the reference and archive groupings much easier.

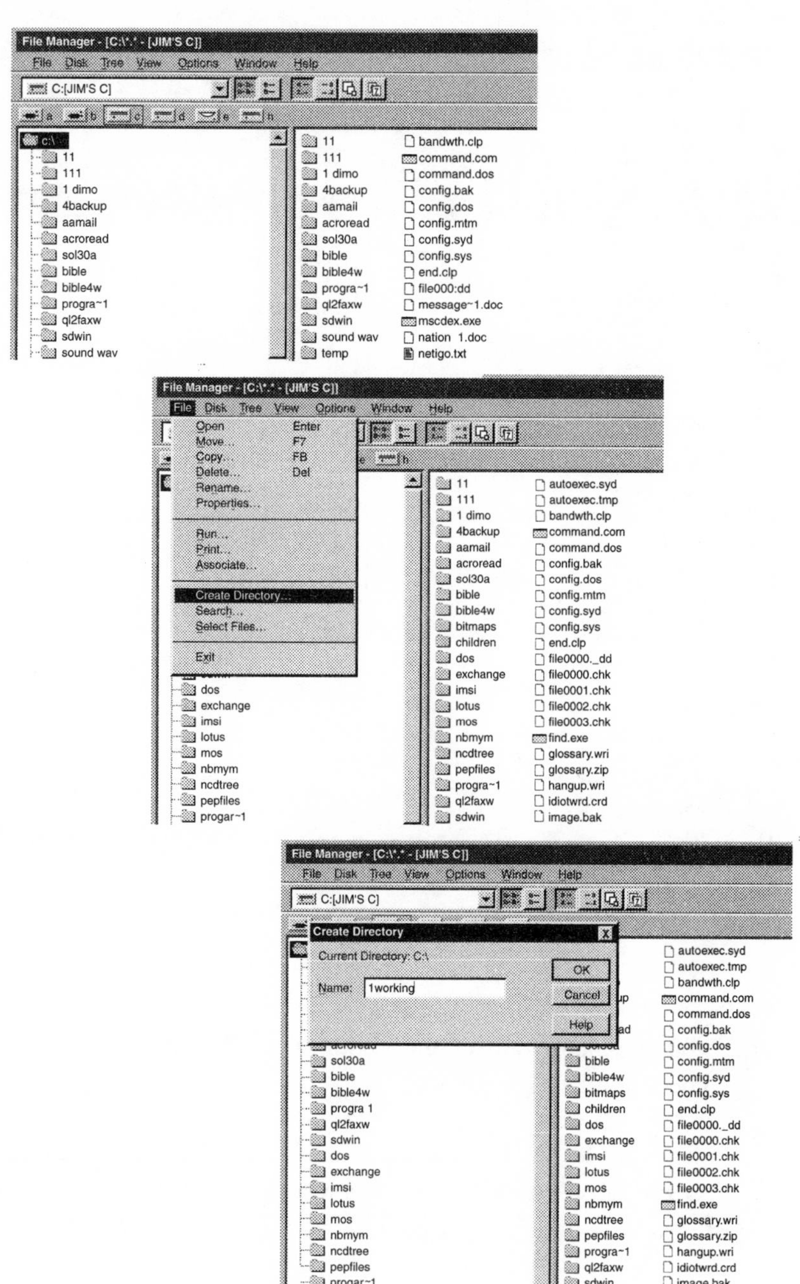

Figure 3.13 Creating directories using File Manager (Win 3.x).

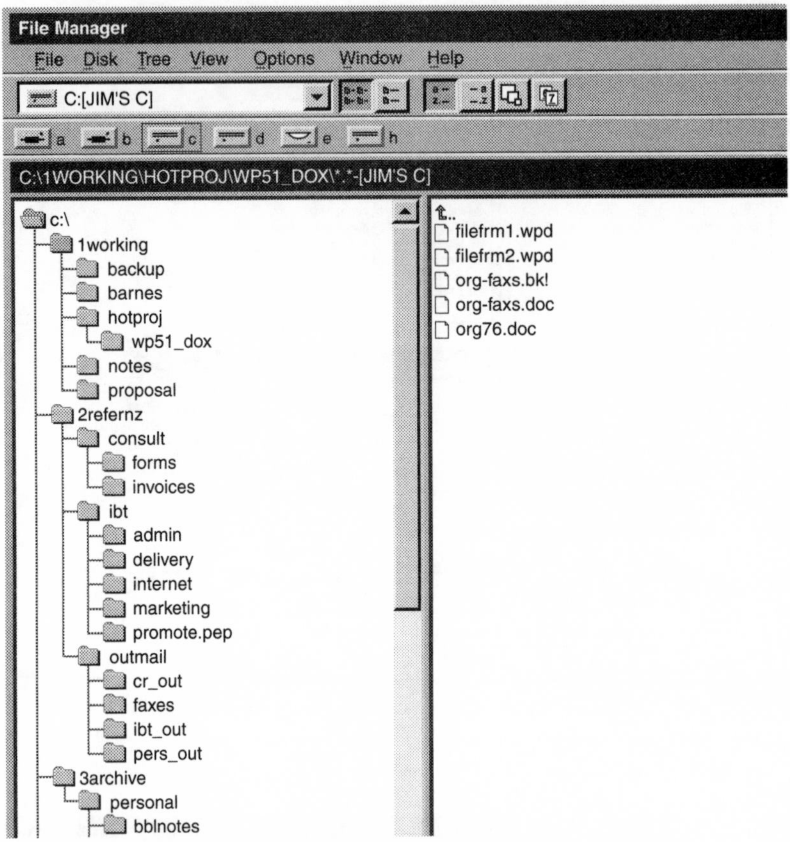

Figure 3.14 Organizing with File Manager (Win 3.x).

Windows 3.x

When you complete your reorganization, your File Manager should look something like Figure 3.14.

For those Windows 3.x users who do not wish to upgrade to Windows 95 but would like to have more characters in a file name, try one of the software utilities on the market (like Office Central) designed to enable you to have more than eight characters in a file name. Also, Norton Desktop

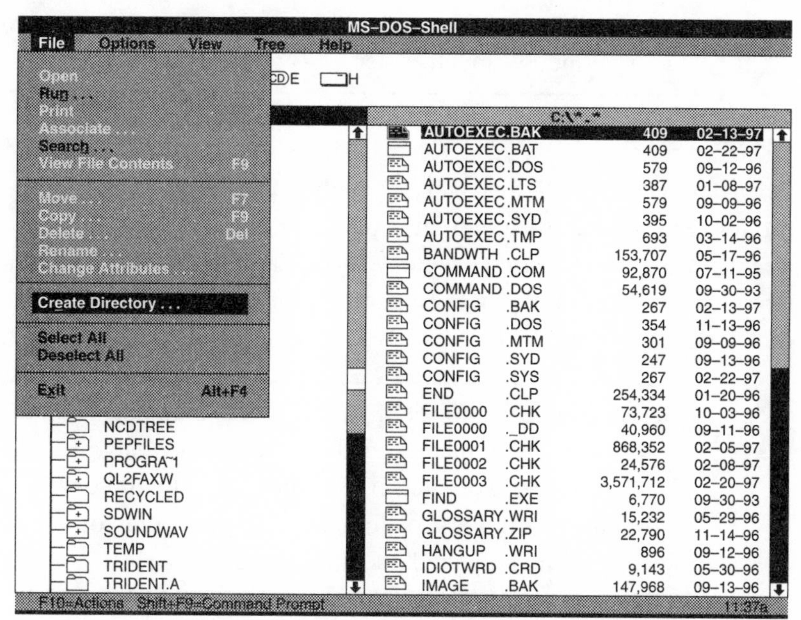

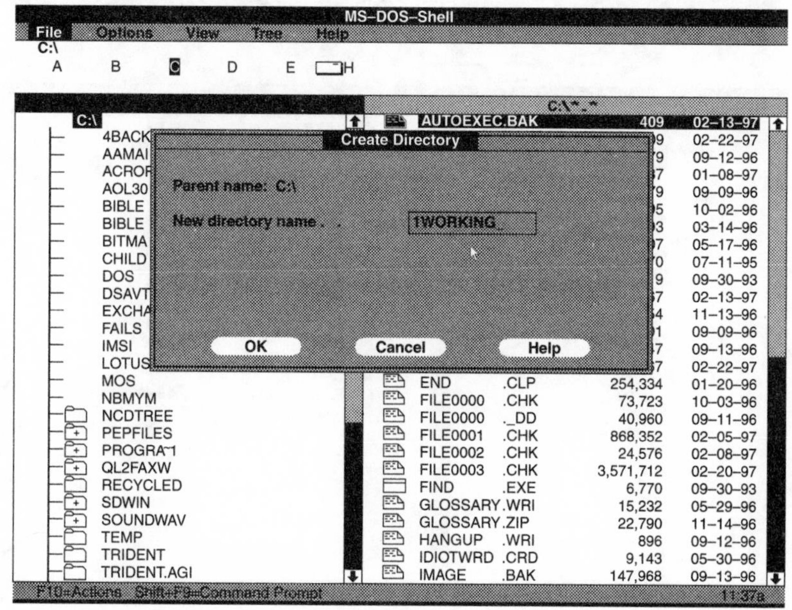

Figure 3.15 Creating directories in DOS Shell.

for Windows provides some nifty features that will make doing Windows much easier.

DOS

Open the **DOS Shell** manager by typing DOSSHELL at a C:\> prompt. Ensure that the highlight is on C:\ at the top of the directory tree. **Choose File, Create Directory. Type the name 1working and press enter.** Repeat the process and create the 2refernz and 3archive directories. See Figure 3.16.

Before attempting to set up directories to organize the hard drive using DOS, make sure that you are operating in a DOS version 5.0 or later. If not—upgrade! It will be well worth the few dollars to eliminate the aggravation of dealing with antiquated operating systems.

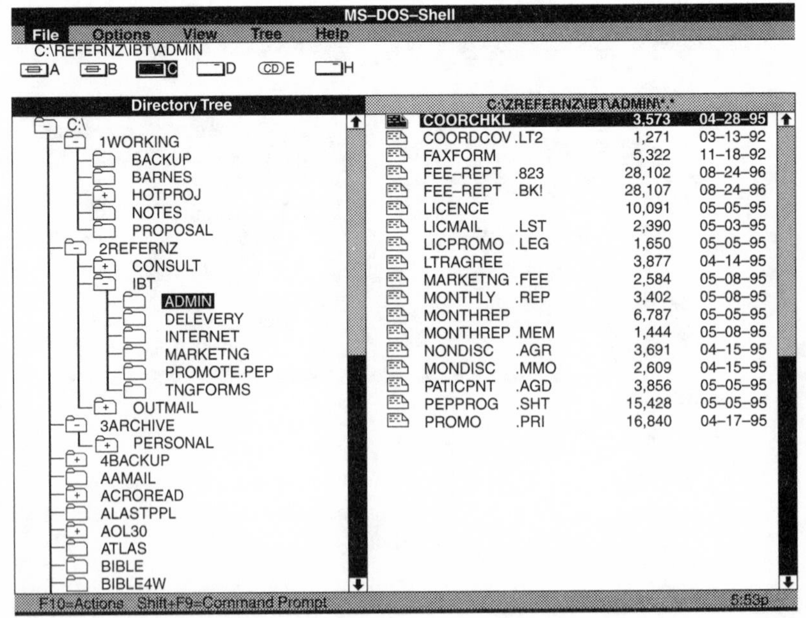

Figure 3.16 Organizing with DOS Shell.

Some very nice and useful programs have been created to enhance the DOS environment. Some of them actually take on a Windows look and feel with colorful icons to start programs and the ability to switch between tasks without closing. I call the process "virtual multitasking." PC Tools and Norton are among many others who not only have such shells, but offer other valuable utility programs in the package.

Macintosh

The instructions for organizing on a Mac operating system (Figure 3.17) are very similar to Windows. If you are a Mac user:

- Create and name three folders 1working, 2reference, and 3archive.
- Using your prepared list of paper files, create a matching set of subdirectories in the 1working, 2reference, and 3archive electronic directories.

TRANSFERRING THOSE DOCUMENTS YOU WISH TO RETAIN INTO THEIR APPROPRIATE FOLDERS

You have the system in place. Now you need to fill it up with your documents. You should have three main objectives in going through the document files of your computer.

1. Purge unnecessary files.
2. Rename, as may be necessary, any document files you keep.
3. Place the document files in the appropriate 1working, 2reference, and 3archive directories. Or as is the case with E-mail and fax documents, in their appropriate folders.

Purging means deleting old and unnecessary files. If you are uncertain whether to eliminate the document, copy it onto a disk for storage. Go into Windows Explorer and bring up your C: drive. Highlight the first application below 3archive and double-click to view the files (on the right side of the screen). Locate the document files for the application.

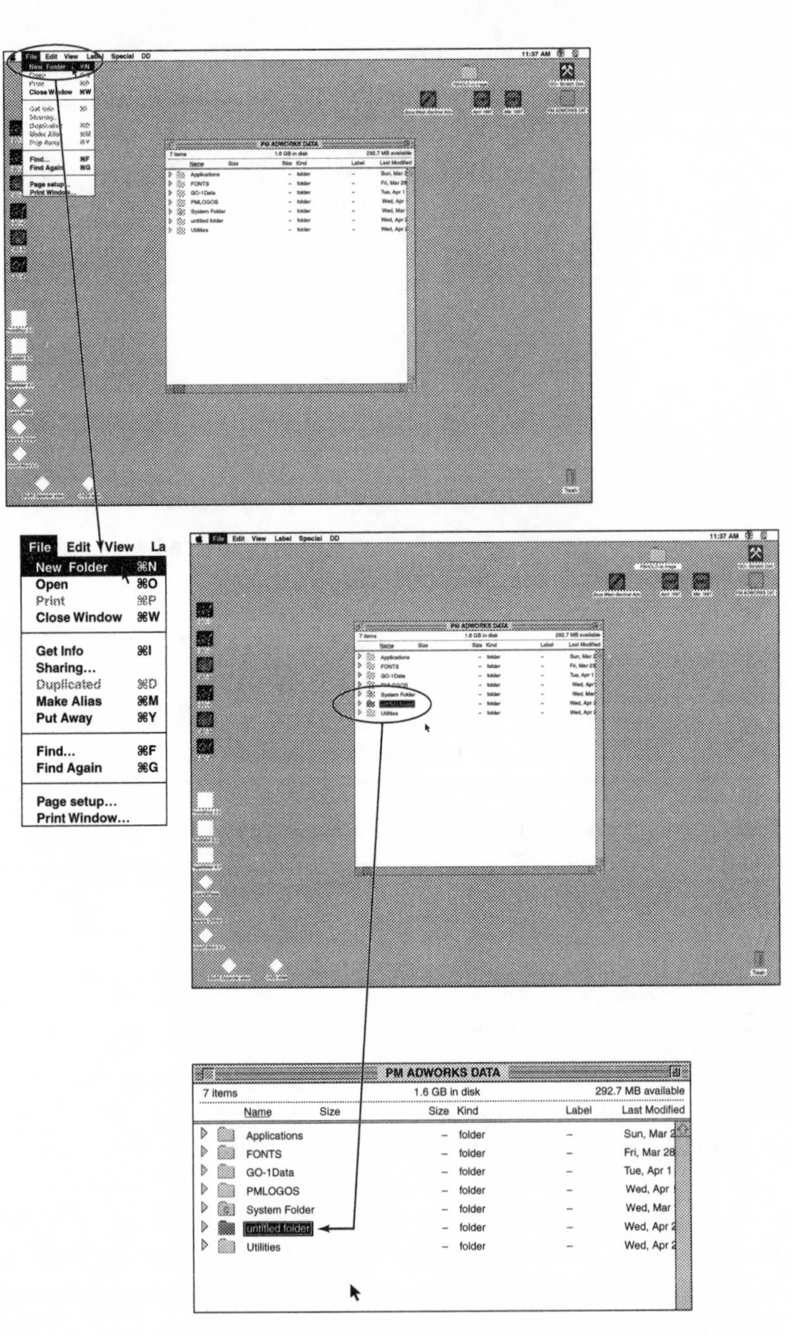

Figure 3.17 Mac hard drive.

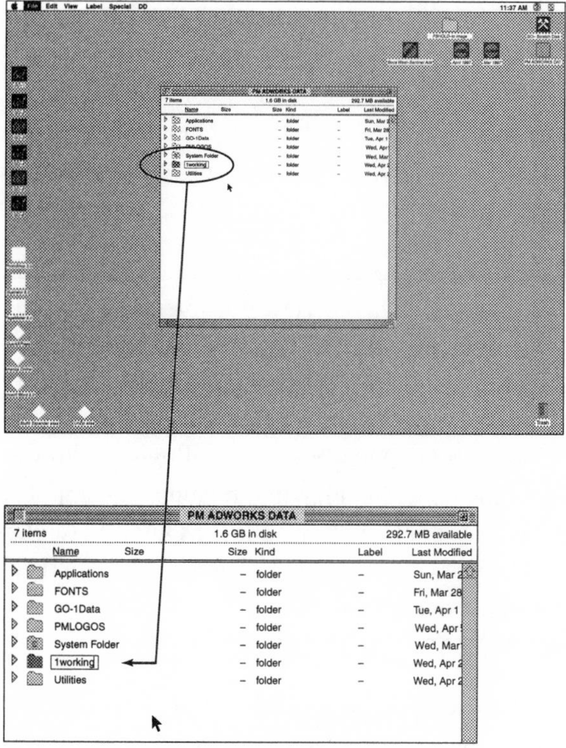

Figure 3.17 Continued.

---

1. If you are unfamiliar with the document's content, view it.
2. Decide whether or not to keep the document file.
3. If you decide to purge the document, highlight the document file and press the delete key.
4. If you decide to keep the document file, verify that the file name is accurate. If you wish to rename the document file, choose **File, Rename and type in the new name and press enter.**
5. If you decide to store the document file on a disk, highlight the document file, put a formatted disk in Drive A: and drag the file to the A: drive. (Each disk should correspond to an archive subfolder. Use the appropriate disk.)

6. If you have decided to keep the document file in either the 1working, 2reference, or 3archive subfolders, drag and drop the document file to the appropriate subfolder.
7. Repeat this procedure with the document files of each of your applications, including (but not limited to) any spreadsheet, graphics, groupware, finance, or other program you are using.

Want a quick way to view documents, even attachments, when you do not have the application normally required to open a document? There is an easy way to do this. I use a software package called Quick View Plus by the Inso Corporation. Quick View Plus allows you to view the documents without having to launch the application. All you do is click once and choose Quick View Plus and the document comes up on the screen. You can then read or print the document.

CREATING A FILE SYSTEM FOR E-MAIL AND FAX MESSAGES

E-mail is arguably one of the most useful tools to come about as a result of the computer. The vast majority of large companies and governmental offices around the world are using E-mail systems to deal with their internal and external communications needs.

Like the paperwork you may have had strewn about the office, E-mail can create chaos if it is subject to poor work habits and disorganization. E-mail, like any form of digital communication or information, needs to be controlled properly. This control begins with a proper organizational system.

Whereas most E-mail messages will be read, answered, and deleted, on occasion there may be a message you want to retain. The first question to ask is, should you be keeping it in the first place? One client told me he was so overwhelmed by the sheer volume of

E-mail messages he got, that he couldn't begin to figure out what to save. Don't fret about this. Most should be deleted. If in doubt, ask yourself the question, "If I needed the information, where could I get it?" If you can easily get it again, delete it.

Unfortunately, many E-mail programs do not let you transfer documents from E-mail to another electronic file. In that case it's possible to cut and paste an E-mail message from one file to another (but that's a multistep process most people are not prepared to do). Another alternative is to print the messages you wish to retain for filing in your paper system. But I do not recommend this because it is wasteful and inefficient.

The best way to organize E-mail is to set up a system of "folders" parallel to those set up on your C: drive for E-mail documents you wish to save. Almost all E-mail applications provide a way to create such a system. Unfortunately, because there are so many different applications on the market, you will have to find out how to create such a system for your application.

For example, John and Philip of the law firm both used Netscape software for Internet E-mail and browsing the Internet. We created an organizational tree (we categorized the messages they would keep and organized folders around these categories) by going to Netscape Mailbox and clicking on File and New Folder. We typed in the name of each folder one at a time and each was added to the Mailbox directory. Each message they wished to keep was dragged from the in box to its appropriate folder.

Faxes Sent Directly to the Computer

Electronic fax applications possess many of the same frustrating limitations as E-mail. Each fax software has a different approach in the design and handling of received faxes. For example, with some software, the user can rename the document and later view that document at will. Other software will not allow the renaming of the document without the document losing the properties allowing it to be opened by the software.

Fax software exists to automatically acknowledge receipt of a fax. For such jobs as letting your customer know an order is being processed, set up an auto-responding fax-back service. It automatically lets your customer know you are on top of the order.

Most fax applications do not permit saving the fax message as its own file and therefore make it impossible to store fax documents in their appropriate folder in the working, reference, or archive directories. Nor do many fax applications allow you to cut and paste within the document. (This mostly has to do with the fact that faxes are pictures of documents, and are not in a digital format. Because of this, the fax uses much more hard drive space and random access memory than your typical digital document.) However, almost any fax application allows you to organize the fax documents in their own set of folders. So the faxes can be organized in their own set of files rather than in the hard drive's working, reference, and archive directories.

Looking for new fax (communications) software? Consider the following specifications for any such software you may choose:

1. It allows for renaming the document file.
2. It allows for editing of the fax document file.
3. It allows the user to store and retrieve documents from locations other than the program directory.
4. It will not interfere with other devices accessing the modem when the fax program is inactive.

Steps to Organize E-mail and Fax Documents

1. Begin by identifying your E-mail and fax system file features.

2. If your applications allow you to transfer fax and E-mail documents to your working, reference, or archive directories, identify those faxes and E-mail messages to save, drag, and drop into their appropriate subfolders.
3. If your applications do not permit transfer of E-mail and fax documents, create within the E-mail and fax application a parallel folder system for all retained E-mail and fax documents. Name the folders based on the categories created for your paper system and C: directories. Drag and drop saved documents into their appropriate folders.

If you are one of those with hundreds or thousands of old E-mail messages, instead of searching through and recategorizing these thousands of messages, choose a cutoff date (like the end of the previous year), create an archive file, mark and transfer these earlier messages to it, and forget about them. If you have to look up an older message, you can do so by accessing the archive file.

To organize documents in Microsoft Fax, double-click on the "Inbox" icon on the Windows 95 work space. You can begin to create your system by choosing File, then New Folder, and typing in the category name. Carry on with this process until you have created folders for all faxes you intend to save. Drag the fax documents to the appropriate folders.

ORGANIZING THE COMPUTER DESKTOP TO ACCESS APPLICATIONS AND FILES EASILY

The desktop is that screen you see when first turning on Windows. It is the electronic equivalent of the top of the desk in your office. It is a relatively simple process to mirror your office desktop in the computer. I highly recommend you do so. Here's how I suggest you do it in Windows 95:

1. Identify your most frequently used programs (e.g., word processing, spreadsheet, personal information manager, etc.). Go into Explorer and find each program on the C: drive. Locate the program application's .exe file. Highlight it and create a shortcut (File, Create Shortcut). Drag the shortcut icon onto the desktop. Do the same for all frequently used programs. Do not worry about where program shortcut icons are located on the desktop for the moment.
2. Create a folder (group icon in Windows 3.1) on the desktop by finding a clear space and click once on the right side of the mouse. Select New and then Folder. Type in "My Programs."
3. Drag the often used program shortcut icons created in step 1 into the My Programs folder. You now have quick and easy access to your most often used applications.
4. Create another folder on the desktop and name it Pending. In it you can temporarily place ongoing documents awaiting a short-term response and access them quickly once the response has been received.
5. For quick access to the printer, click on the Start menu and choose Settings, Printers. Drag the printer icon onto the desktop. Then simply drag any document you wish to print to the printer icon on the desktop and voilà, a printed document.
6. I personally prefer a clean and spacious desktop. Therefore I have arranged the icons along the edges of the desktop. To neaten them up, click on a blank space on the desktop and choose Line Up Icons.

Your desktop might look like Figure 3.18.

Desktop Organizing Tools

Organizing your computer desktop can be easier to do if you use a Windows organizing tool such as Norton Desktop for Windows. The idea is to simply create a group that contains all of your work programs (or those you frequently use). Norton permits you to do

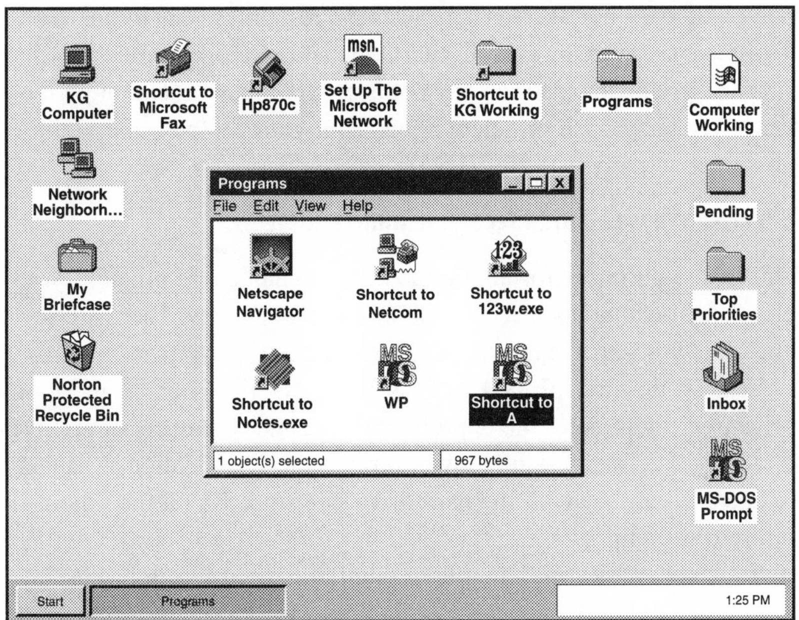

Figure 3.18 Desktop (Win 95).

some things that I could not do in plain Windows 3.x such as drag a file onto the desktop. For example, if you are working on a particular project or document and need to open the file frequently, you can do so very quickly by dragging the file's icon and dropping it on the icon for the program or viewer used to view the file. This process is made possible by the feature which allows you to place program icons directly onto the desktop. You can also open such a file by dragging it to the icon of the application in which it was created.

With or without a desktop management program, the idea is to simplify your electronic desktop in much the same way that you would simplify your office desktop. The basic Windows 3.1 program allows you to create categories of groups to make it easier for you to find the program you need. Having a program appear in more than one group does not decrease the system's resources. You might therefore have a program (say Lotus Organizer) in your toolbox, your start-up group, and also in your Lotus group. Please note that periodically you may wish to delete items from your toolbox that you are no longer actively using.

Effective Ways to Find Electronic Information

One of the purposes for taking the effort to name and organize your electronic documents is to be able to find them easily. There are several tools you can use to find the files you need.

Usually the first place I look is Windows Explorer, at the hard drive tree. I will enter the appropriate working, reference, or archive directory and trace the file down through the organization. I happen to be very familiar with the hard drive tree since I took the time and effort to set it up.

If I cannot easily find the file through this type of search, I use the Find function in Windows 95. This is as easy as entering the Windows Explorer, selecting Tools, Find, and Files or Folders, and typing in the file one wishes to find; Windows will then bring up any files with that name. Double-click on the file and Windows launches the application and brings the file up on your desktop.

I also use a software package called AltaVista Search My Computer from Digital Equipment Corporation. AltaVista Search is the personal version of one of the Internet's most powerful search engines, AltaVista. When installed, it indexes all files and automatically updates its indexes daily. You search for files and text in the files through your browser software (in my case Netscape) in the same way it searches for terms, files, and information on the Internet. Type out a phrase and AltaVista searches all files for use of the phrase and up pops a clickable link to the file. This is a useful tool when you know what file information you are searching for, but do not remember the name of the file.

There are several such search software applications on the market. Besides AltaVista Search My Computer is another product called Hyperlinks. Check out your computer store or the Internet for the one best suited to your needs. I highly recommend doing so.

PROTECTING YOUR ELECTRONIC INFORMATION

Backing Up Your Electronic Documents

Make it a practice to regularly back up your files. As part of the research for this book, I prepared a survey on the topic and sent it out by mail and posted it on our company Web site. One survey respondent never backed up the files on his notebook hard drive. His computer recently crashed and he lost all of his files. He didn't even have what's referred to as a boot disk to turn his computer on! This is a common occurrence. What made this story unique is that the person is a computer teacher! As he said, "This is bad!"

If backing up onto floppy disks, follow the rule "one subject, one diskette." It is surprising to see how many people try to economize with disk space when it takes ages to review and find files on a disk. A disk might cost a dollar. You have to ask yourself what your time is worth.

Personally, I am a bit paranoid about losing my computer files. As a result, I use a tape backup system (Colorado Jumbo External Tape System). I find it much easier to use than the backup provided by the operating system. The Windows backup program calls for disk backup, which can be quite time-consuming. As a disk fills up, you must remove and replace it in your A: drive. Since most disks are 1.44 megabytes, any normal hard drive backup will consume many disks. But with Colorado, I just pop in a 500-megabyte tape and click on Full Backup. The system takes over and doesn't require you to be present unless you have more than 500 megabytes of information (and even then, it's as easy as putting in an additional tape and clicking Continue). Better yet, the software allows you to schedule automatic backups (daily, weekly, etc.).

Colorado is only one of many backup systems available on the market. One client uses an external Iomega zip drive that plugs into the parallel printer port. The computer treats it as an external drive. The Ditto zip drive (one of the several models of zip drives from Iomega) holds up to 2 GB (gigabytes) of information. All the client has to do is drag the document files (following our suggestion of putting all document files in a working folder [directory]) over to the zip drive icon and all of the working files are backed up onto one disk. It takes about four to five minutes to back up 30 MB (megabytes) of information.

If you don't want to hassle with drives or disks for backups, you might look into a backup service offered by some of the larger telephone companies. Connect your computer to their server, and you can transfer a copy of your files to their secure off-site storage computer. One such service is workMCI Backup.

Utilities

Utilities are software applications that, among other things, detect and solve computer problems. There are utilities designed to detect and prevent viruses from entering your computer system, recover files and deal with a computer crash, and enhance the features on your operating system.

I highly recommend you get Norton Utilities by Symantic Corporation. Norton helps prevent computer crashes. Computer crashes are not common, but when they occur (another Murphy's law—they *will* occur!), you'll be grateful for having a utility that makes it easy to recover your files and get your system up and running again.

Try "exercising" your computer periodically. Simple programs like Memmaker and Defrag maximize your computer's memory capability and keep your hard drive functioning optimumly. Norton Utilities "oils" your computer and ensures it will last with a minimum amount of trouble.

Norton monitors a hard disk's integrity and notifies you as soon as it detects potential problems. If your computer will not start, Norton can start it. With Norton, you can locate files that are candidates for deletion, compression, or being moved either because you no longer need them or because they take up more space than they are worth. Norton ensures your hard disk is organized for optimum performance, notifies you if it is not, and corrects the problem.

Take my word for it. You need such a utilities package.

MANAGING IT ALL

You will discover that having organized your tools as described in Chapter 2 and Chapter 3 is going to make processing information and work much easier to cope with. Even if you change nothing else, you'll work more efficiently. But I would suggest that you'll be even more productive if you polish up your work practices as follows:

1. Identify all channels of information and work (company E-mail, postal mail, Internet E-mail, faxes, in box, voice mail, meetings, etc.).
2. Introduce routines to periodically check each of these (postal mail may be only once per day, E-mail and voice mail two or three times per day, etc.) and process them completely. If you forward an E-mail message to another, schedule a reminder to follow up. White-collar productivity shrivels if you do not act now. Acting now loses its effectiveness if you do not follow through.
3. If the task gets shoved in your face, *do it now*. If it is impossible to get it done immediately, schedule a completion time in your calendar. Don't fool yourself. Most things can be done the first time you come in contact with them (or if not completed, often at least the next step or two can be done, moving the task closer to completion).
4. Continuously evaluate the information you process. Should you be getting it in the first place? Regularly eliminate, at the source, all nonessential information.
5. Delegate liberally. Be clever about using the resources around you to get things done. If you delegate, follow up to ensure it is done.

6. Organize as you work. If you create a new document, place it in the correct (working, reference, or archive) directory (folder) immediately. Organize your materials. Perform the job in an orderly fashion. Clean up afterward.

SUMMARY

Gaining control over the work generated by the computer begins with creating an organizing system for all the electronic information you receive. Using the paper-based system you have set up in your office, create a mirror system on the computer. Do this with all document files as well as E-mail and fax messages you keep. Once you have a place to put things, make it part of the work cycle as described in 4S, and clean up afterward! You have organized your computer. Now, let's go to Chapter 4 to see how the computer can organize you.

FOLLOW-UP ACTION POINTS

✔ **Organize the documents on your computer**, beginning with a full backup of your computer hard drive.

✔ **Create three folders** (directories) on the C: drive. Name them 1working, 2reference, and 3archive (the numbers move the directories to the top of the C: tree).

✔ For Windows 95 users, **create a shortcut** from the 1working folder to the 2reference folder and a shortcut from 1working folder to the 3archive folder.

✔ Using your prepared list of paper files, **create a matching set of subfolders** in the 1working, 2reference, and 3archive electronic folders.

✔ Go through each of your application document files and **transfer the documents to the appropriate subfolders.** View the documents you are transferring and delete any unnecessary files.

- ✔ Direct the applications to **search for the document files in the 1working directory.**
- ✔ **Create folders for important E-mail messages** you wish to save. Folders should be categorized to match those on the C: drive and your paper-based file system. Move E-mail messages you wish to save from the in box to the appropriate folders.
- ✔ **Create folders for important faxes** you wish to save. Again, categorize folders in the same way as your paper-based system. Move faxes you wish to save from the in box to the appropriate folders.
- ✔ **Identify all channels of information and work** (company E-mail, postal mail, Internet E-mail, faxes, in box, voice mail, meetings, etc.).
- ✔ **Introduce routines to periodically check each of these** (postal mail may be only once per day, E-mail and voice mail two or three times per day, etc.) and process them completely. If you forward an E-mail message to another, schedule a reminder to follow up. Productivity shrivels away if you do not act now and then follow through.
- ✔ **Continuously evaluate the information you process.** Should you be getting it in the first place? Regularly eliminate, at the source, all nonessential information.
- ✔ **Organize as you work.** If you create a new document, place it in the correct (working, reference, or archive) directory (folder) immediately. And stick to the motto: "Organize your materials. Perform the job in an orderly fashion. Clean up afterward."

PREVIEW

Time and money. Study after study shows the two resources we never seem to have enough of are time and money. How do you get either under control? Account for it! The computer makes the complex activities associated with these resources much easier to deal with. In this chapter you will learn:

☞ How to use a personal information manager (PIM) to multiply your personal productivity.

☞ How to use finance software and electronic banking services to get all aspects of your finances under control.

4

Organizing Yourself with a Computer: Time and Money

Congratulations! You have organized your desk and computer. You are better prepared to *act*. But act on what? How can your newly organized electronic tools help you to get more of the important things done? How can these tools give you more control over two of your most precious resources—time and money? How can your electronic tools automate work processes and make it, once again, *easier* to act on and get things done?

There are thousands of applications designed to make the most of a computer. Depending on the nature of your work, specific applications help you achieve specific goals. But there are two general-purpose uses of the computer that, in my view, make it an indispensable tool:

- Organizing your day-to-day tasks.
- Accounting for and controlling your finances.

Here's how.

ORGANIZING YOUR PLANS AND TASKS

I believe people are far more likely to do things that are *easy* to do. If you want to get more done, the computer is a wonderful tool, because it makes it easier to organize what needs to get done. The computer is particularly useful in tracking things to do and providing access to the vital information necessary to get things done. The software used to organize your day-to-day activities and tasks is called a personal information manager (PIM), a term coined by Lotus Development Corporation when it released a software program called Agenda. The PIM category encompasses names, dates, addresses, phone numbers, notes, lists, text, financial information, and whatever else anybody decides should belong. Ever since Lotus coined PIM for Agenda, the category name has stuck, and there are now scores of PIM software applications on the market.

In order to be able to see the potential benefits of organizing your plans and tasks with a computer (Figure 4.1), it's helpful to identify the problems or difficulties that people run into in the course of doing their day-to-day work.

Figure 4.1 Before and after organizing yourself with a computer.

CHALLENGES AT WORK

Keeping Track of Your Tasks

To get things done you have to first identify what it is that you need to do. Depending upon the number of people with whom you interact and the different hats you wear, both professionally and personally, it is not uncommon for us to have several thousand tasks open at any given time in our lives. Thousands of tasks are simply too many for us to leave to memory. Unless these tasks are organized in some way, they become overwhelming.

Want a low-tech way to keep track of all you must do? Get a simple log book (a good type is a school composition book) and note down everything you need to do in it, rather than on separate pieces of paper or Post-it notes.

Overcoming Procrastination

We all have a tendency to put certain things off, especially tasks we find unpleasant, boring, complicated, or difficult. And we use any excuse we can find to postpone doing these types of tasks. For example, if we do not like speaking to an individual, having trouble locating the telephone number is a wonderful excuse not to call. Conversely, if we are organized to easily execute tasks, we are far more likely to do them even if they are unpleasant. One of the big challenges we face in our work is devising schemes and systems that encourage us to act and overcome this tendency to procrastinate.

To overcome tendencies to procrastinate, try doing the unpleasant tasks first.

Dealing with Today's Overwhelming Flow of Information

Digital technology has provided us with ways to get things done more quickly and efficiently. It has also significantly increased the number of ways information and work comes our way. You have the fax, both manual and electronic, sending instant communications. You have

E-mail (which might be located on your internal communications system as well as the Internet). Meetings haven't gone away (though many of us wish they would!). You still have the desk telephone. In addition, you have the mobile telephone and pagers. Any one of these channels of communication can have vital information or work to process. Add to this the fact that many companies have been downsizing and eliminating resources. Now you have a situation with more things to do, more channels to keep on top of, and fewer resources to utilize!

One way of coping with information overload is to stop it from getting to you in the first place. Analyze the flow of information with the idea of eliminating unwanted and unnecessary communications at the source.

Keeping Up with Technical Developments

Technology is wreaking havoc in industry after industry. An artist friend of mine who made a good livelihood illustrating posters for the movie industry has phased out the business and turned to painting. He had to decide between his love for hands-on painting and learning how to create the poster illustrations on a computer. It might have been a difficult personal choice, but he had witnessed turmoil in his and related industries that helped him make up his mind about his future course. My point in bringing this up is not to say whether or not my friend made a "right choice," but rather that technological changes are going to happen and you have to be prepared to accept them. Certainly, you have to keep a keen eye out for change in your own profession. Most of us do not have the luxury of closing down our business and moving to a new career.

Getting Others to Get Things Done

The work problems so far noted can, to one degree or another, be dealt with through our own efforts. If we process things immediately and don't put them off, we can get more done. If we organize ourselves well we can be more efficient and effective. However, too often our problems in work are not of our own making. Our work problems may be the result of others not doing their jobs. Getting others to do what we intend them to do is very important. A certain portion of our day will be consumed following through with others to get things done so we can complete our own work. Delegating tasks and having good people skills do make a difference. Yet the largest single factor in getting others to get things done is your persistence and follow-through. Having a good follow-through system in place is essential to successful execution of work.

Revising Wasteful Policies

It's one thing to be well organized, to know where things are, to have your tasks simply laid out, and to do your job in an effective way. But no matter how efficient or effective you may be, you will be impacted by systems or processes within the organization that are themselves wasteful. An example of this is having to attend poorly planned meetings. Too often meetings produce little result, yet you have to waste your time participating in them. Or, you might be very busy executing a series of actions that seem reasonable but in the end produce very little of value. We often see this in government. Rules and regulations brought into existence in some bygone era to solve a problem of the past are still in force and create problems for you today.

I see the waste of time, money, and energy in companies all the time. In discussing the matter with senior executives who should and could do something about it, I usually find that the problem boils down to not enough time, energy, or focus on the executives' part to deal with and solve the problems. If the executives were well organized and had effective systems in place to be able to keep on top of their tasks and functions, they would be able to solve their policy and structural issues.

HOW TO HANDLE TIME—THE PERSONAL INFORMATION MANAGER (PIM)

Good work habits go a long way toward resolving time-related issues (see Chapter 1). You have another resource: a good calendar system (Figure 4.2). An effective calendar system helps you to:

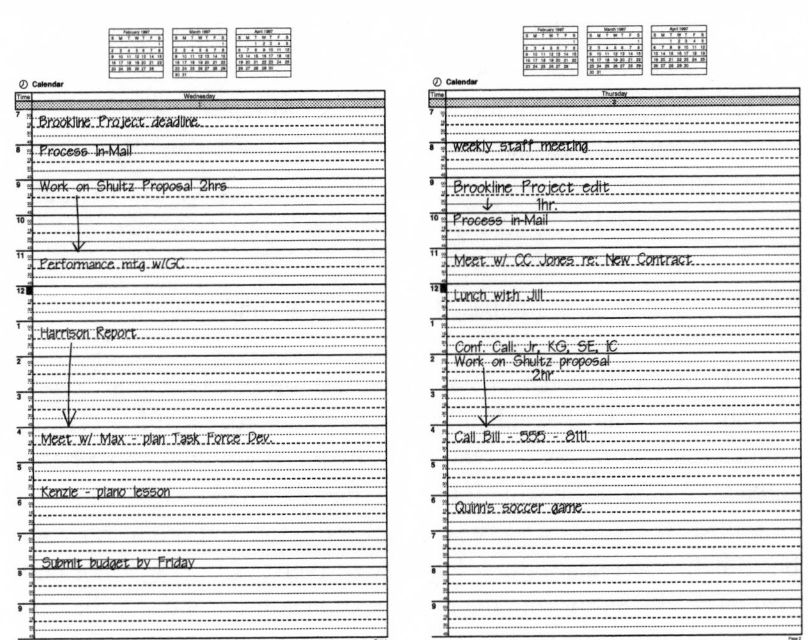

Figure 4.2 Calendar system.

1. Remind yourself to do things in the future.
2. Note appointments.
3. Write to-do lists, or plan for the upcoming week.
4. Note important deadlines.
5. Work back from deadlines and note down milestones.
6. Remind yourself of recurring events such as birthdays, holidays, anniversaries, and other special dates in your life.
7. Write notes from meetings.
8. Keep address and telephone information.
9. Provide general information, such as: time zones, area codes, and postal codes.
10. Block out time for work or recurring activities such as scheduled time to meet with your employees on a weekly

basis or schedule time to process your E-mail, paperwork, paper mail, and so on.

11. Keep personal information such as insurance numbers, driver's license number, and auto registration number.
12. Note personal medical information, such as the date of your most recent physical, your allergies, health insurance, and so forth.
13. Organize your life based on your goals and values.

A computer—whether a stationary desktop one in the office or a notebook or smaller handheld model—with a robust PIM software application automates a great many of the traditional paper-based calendar functions, making it that much easier to use.

A PIM won't solve every work-related issue you might run into, but it helps you deal with a majority of them. A PIM will not only deal more efficiently with the aforementioned tasks, it can also:

- Allow you to view (and if you wish, print out) the information in your calendar in many more ways than are possible in a paper-based system. Most PIMs allow you to view the calendar by day, week, or month as well as by assigned category. These options make the information much more useful to you.
- Automatically copy input information to many places (calendar, to-do list, person responsible, project, etc.).
- Set alarms which will remind you of things you need to do (if you are away from your computer it can send a text message to your pager).
- Connect to the Internet or your company's local area network (LAN) or intranet and send and receive E-mail.
- Notify you when you receive E-mail.
- Schedule group meetings by automatically finding open time slots for meeting participants.
- Attach electronic documents to messages (and group scheduled meetings).
- Automatically schedule recurring and multiday events.

- Automatically roll over incomplete to-do items to the next day.
- Find information easily with PIM search functions.
- Link tasks in such a way that you can connect several items' dates, so that when you change one, others move relative to it.

It's Not That Easy to Use a PIM

I do not wish to give the impression that converting to an electronic information manager will be an easy process. Many people I know have tried and abandoned their PIMs.

However, few of those who abandoned their PIMs used their paper systems very well to begin with. Their paper calendar may have been used to write a reminder of an appointment but seldom went further than that. Their calendar system was not a *planning* tool. It was a *critical* item reminder system. If they didn't bother to write things down in their paper-based system, they didn't bother to write things down in their PIM.

It takes quite a bit of work to convert to an electronic reminder system. It takes effort to choose the best electronic system. One has to learn the software (and sometimes hardware if new to using a computer or if one has purchased a palmtop or personal digital assistant). There is the transfer of information from the paper-based system to the electronic system, a time-consuming process.

One of the common troubles using a PIM is the restrictions placed on you by having your calendar and planner on your desktop. The problem is you are not always sitting at your desk (you may be on an errand, at a meeting, or at home) when a reminder needs to be added to the PIM. It is also sometimes easier to open a paper calendar system than, say, a computer not left on.

Probably the biggest difficulty to overcome is the tendency to fall back to our old ways of doing things. If we are used to behaving a certain way (and may have been behaving that way daily for years), it is very easy to give up and fall back to old ways we have found comfortable. It is sticking it out until you have created a new behavior that is the biggest challenge.

Converting to a PIM and overcoming these very real difficulties is not for the weak-hearted. You must be determined to make it work.

Few things of value come without commensurate effort. I promise you this: If you can manage to overcome the difficulties you encounter, you will be richly rewarded. You see, it is the attention to the details of work and systems in place that make it easier to pay attention to the details that monitor the control you exert in work and life.

Typical PIM Uses

A few real examples illustrate how valuable a PIM can be:

Kathryn is head of office services in a large insurance company. She uses a PIM called DayTimer Online. She has 28 people reporting to her. These employees are spread over 7 floors of an 18-floor building. Each morning when she gets into the office she updates her daily calendar on the computer to include all of her to-do's and appointments for the day. She uses this time to delegate any tasks from the day before through the use of E-mail. She prioritizes her tasks, "one, two, and three." First are tasks "I would be fired if I didn't do!" Second are vital and important tasks to get done, but not as critical. Third are the tasks "I would like to get done but do not normally end up doing."

Because her employees are spread out over so many floors, she has quite a bit of traveling to do to see them. To save time she prepares a daily follow-up list for each person on her staff. She prints these lists out along with her full daily calendar with all her to-dos and puts them on a clipboard she takes with her from floor to floor. (She prefers to write the information down on paper rather than type into a small handheld organizer.)

As she explains it, the key benefit for her is that she doesn't have to obsess about forgetting things; it's all right there in front of her.

Dagfinn is the managing director of a trade association for independent tanker owners. He uses a popular PIM called Microsoft Schedule+. It's a groupware application that allows anyone in the office to access his schedule directly. Since his clients are spread over 41 countries, he travels regularly. When he's away he can connect to his home-office server for E-mail and staff can input new appointments for him. Upon his return, he can transfer all the changes from his notebook to his desktop computer.

Each morning Dagfinn reviews all his tasks in Microsoft Schedule+ and enters any new ones into the calendar. All of his tasks are date-related. Anything not done from the day before is automatically carried over to the next day. In this way he reviews new appointments and plans out his day. He has a clear idea of his tasks for the day.

Anders is the managing director of a thriving branch of our Institute for Business Technology consulting firm. The PIM that he uses is Lotus Organizer, customized for both his and his consultants' needs. The Organizer allows for commonly shared calendars and files. Their customized Organizer files have been designed to keep a master client address and phone list, pending items, and notes from calls with these clients. Active clients are labeled with the consultant's initials and inactive clients are so noted. A consultant who gets a new lead can look in the common file to see if anyone is working on the prospect and if there is any existing information about the lead.

Each consultant keeps an individual Organizer file to track personal tasks, reminders, and follow-ups. The consultants add pages to their own Organizer for projects they are responsible for. Project tasks show up in their calendars automatically. Meanwhile, Anders keeps his own tasks and follow-up notes in his Organizer calendar. The Organizer also provides Anders with a list of people the consultants will be calling so he can keep apprised of what's going on. Between these two lists, he can view his daily calendar and clearly see everything to be done that day in business.

One of the niftiest features of the Organizer for Anders is its signal function. When Anders has an appointment, he has the Organizer signal him with sufficient time in advance to get to the appointment. If he is trying to contact a client by telephone and the client cannot take the call, Anders will ask for a time when the client will be available, and put a signal in the Organizer to call the client back at the agreed-upon time.

John, our lawyer friend, also uses Lotus Organizer. For John (as is the case with most of us), time is money. For billing purposes it is essential that he keeps an accurate record of the time spent on the work he does. He uses his Organizer as a record for all of his billable time. He does this by scheduling "appointments" with himself to do specific work tasks. When new work comes up he notes it in his calendar as an appointment in the future. He keeps his electronic calendar "open" at all times and marks events as they occur. His secretary can access his Organizer, print a copy of his calendar, and input it into the central system of the law firm. For easy access, John maintains a complete client listing in the Organizer with basic client information. And he uses its anniversary function liberally.

The benefits from his PIM: First, he can forget about the work he's got to do in the future because he knows it's scheduled and when the time comes he will do it. Secondly, he's made it easy to get the work done, because all the information he needs is immediately accessible.

Choosing a PIM

There are many PIMs on the market from which to choose. Most will have unique features designed for special applications. Some are so feature-rich they provide for almost any application. Microsoft Schedule+ and Lotus Organizer are two such feature-rich PIMs. Both, however, hog random-access memory (RAM) and hard drive space, require pretty powerful computers to run, and may not be better than PIMs designed for special applications. It is only reasonable then to take time to find the right PIM for you. The first step is to identify the features that best match the nature of your work and what you are trying to accomplish. For example, if you are a sales person you would want a PIM that has a versatile contact manager feature. If you are a manager, you might want to have a PIM with project management and group scheduling features.

After you have identified the features that best fit the nature of your job, you will want to make sure your PIM will reinforce any system you put into place to execute your job functions.

Take me as a case in point. I run a business that franchises services and products in 20 different countries, and I service these country franchises. I am also a writer and public speaker. I have customers to answer to. I have a network to manage. And like most businessmen, I am a dad and responsible for all that comes with the dad role. To keep on top of it all, I use a PIM called Lotus Agenda. Although a DOS application not made or supported by Lotus any longer, I find it the best PIM for me. In Agenda I:

- Write everything I need to remember to do and keep on top of.
- Keep important information including business plans, address book, yearly events, schedules, deadlines, calendaring items, meetings, reference material, loans and mortgage information, auto, health, life, disability, and home insurance information, my children's passport and Social Security numbers, banking information, directions, restaurants I frequent, and airline frequent-flyer information.
- Keep private information—from my retirement investments and financial information to family and home matters.
- Have approximately 20 projects that I am running at any one time. Each project may have anywhere from 10 to 200 tasks involved. The vast majority of tasks are date-related, so they will show up in my calendar at the appropriate time for me to act on them.
- Have noted my goals and objectives and the project plans to execute them, as well as any reminders. I can quickly scroll through any of these and transfer tasks to be done in the next week to my weekly plan.
- Have a good contact management application and, therefore, I am able to quickly go through and determine with each of my clients what I must do in the upcoming week and assign it to my weekly plan.

I especially like Agenda's feature that enables me to type a task and Agenda automatically cross-references it in already established categories. I can type the task "call Bill Smith on Friday about Project X." Agenda, being text-sensitive, automatically puts the task into my calendar for Friday, and puts the same reminder under Bill Smith's name and also under the project called Project X. It is very difficult indeed to lose sight of the task when a reminder of it is automatically placed in every conceivable location.

There are many PIM applications with similar features. Again, a little homework will help you locate the one best suited to your needs.

PIMs Available Today

A variety of PIMs are on the market, including Action Plus, Ascend, DayRunner, DayTimer, ECCO Pro, and Microsoft's Schedule+, to name a few. To give some idea of the different features of PIMs available today, the following are brief feature summaries gleaned from reviews in magazines *PC Magazine* and *PC Week*:

Action Plus offers basic contact management features like the creation of recurring appointments. It also takes sales management a step further by including a sales module that lets you track inventory and sales.

Ascend allows you to enter ongoing tasks directly in its Master Task List, or drag them over from the prioritized Daily Task List. Ascend handles random information through its Journal and Red Tabs.

DayRunner's visual presentation is clear and easy to understand, although it's not an information management powerhouse like Schedule+. Its functions are simplified because choices are limited.

DayTimer's calendars, its strong suit, allow you to jump from daily to weekly to monthly views by clicking on the appropriate icons in the upper lefthand corner. A double click on the calendar will bring you to a dialogue box that lets you specify information.

ECCO Pro gives you fingertip access to every appointment, task, project, conversation, or action time that is important to you. And with ECCO's group scheduling, file sharing, and synchronization, you can extend this productivity to your work group.

Schedule+, with the calendar and the planner window, lets you easily coordinate schedules with other people, and the Meeting Wizard helps you cut through the complications of setting up a workgroup meeting.

Getting Started with a PIM

Having assisted scores of professionals to convert from their paper-based calendar system to an electronic PIM, I offer here some hard-won suggestions based on my experience.

1. Choose the PIM suited to your needs.
   a. Does your work bind you to the desk?
   b. Do you intend to use a PDA or handheld computer as a companion to your computer? If so, ensure your PIM operates in both environments, and synchronization is easy.
   c. Does your work require lots of people contact and follow-up? If so, you might consider a rich contact manager application like Act, which has its own calendaring function. Or you might choose ECCO Pro, a PIM that has a more complete contact management module than, say, Schedule+ by Microsoft.
   d. For a PIM with all of these features, you might consider the Lotus Organizer. With Lotus Organizer, you can use PEP Planner, a special application designed to reinforce the principles of PEP covered in this book.
2. Once you have chosen and installed your PIM, begin by reviewing the help section. Skim through the "How do I" help lists, found in almost all PIM applications.
3. Begin entering information in the calendar section of the PIM. Note appointments and reminders in the calendar views. Copy recurring events, appointments, deadlines, and so on from your existing calendar. Do not attempt to maintain both an old and a new system; too many items will fall through the cracks. Our experience has shown that if you are not committed to using a new system, you will slide back into using the old paper system with which you are comfortable. Allow using the PIM to become a habit.
4. If you have an existing electronic PIM and the information is importable into the new PIM, your job is much easier. Transfer the information.
5. Set up the address book in the PIM for managing contacts.
6. Begin using the to-do lists function for reminders.

7. As you run across important information (a new credit card number, for instance), use the PIM to store it. Add to the PIM as new information pops up. You will see it fill up quickly.
8. Learn the more advanced features of your PIM.
9. Set up the PIM to work according to your preferences. Most PIMs give user options on how certain operations will function. Discover what these options are and see how you might format the PIM to better fit your methods of working. Use the trial-and-error method to determine what works best for you in the design of your PIM.
10. Set your computer's start-up function to bring up your PIM when first turning on the computer. The PIM becomes the first thing you see when turning your computer on, enabling you to review the upcoming events of the day.

Start!

PERSONAL DIGITAL ASSISTANT (PDA)

PDAs refer to handheld electronic tools that organize information in much the same way as a PIM. In fact, many of the PIM software packages will be installed on a PDA. A PDA is usually small enough to put in your pocket and can be used to track information and even in some cases as a means of communication (sending electronic faxes and E-mail). PDAs will be covered in Chapter 5.

CONTROLLING YOUR FINANCES

You can account for all of your tasks and time with a good PIM. Now comes money! Using a computer to gain control of your finances is one of the most helpful features of new technology. Identifying where your money goes is very similar to identifying where your time goes. The remedy? Account for it!

By converting from a paper-based finance system to an automated system with a software package like Intuit's Quicken, you will quickly begin to increase control over your finances and speed up the processing of your financial work. Complement this with what's called on-line banking and you will eliminate a great deal of time, effort, and stress dealing with your finances. Let me show you how it works.

I will cover the use of Quicken because it is the most popular personal finance application around. Like PIMs, most of the finance applications on the market offer similar features. Seeing how one works will give you a good idea of how most will function.

Quicken has helped me in the following ways:

- No longer are there any addition or subtraction errors in the checkbook register.
- Check writing is very fast. Quicken has a feature where it remembers previous checks you have written and automatically fills in the check as you type the first few letters of the payee. Since I write most of the checks on a bimonthly basis and the majority are repeat payees, this saves a great deal of time. I would estimate that it takes one-fourth the time it used to take.
- The reports Quicken can create are very beneficial in detailing whatever you need to know regarding income, expenditures, transaction details, and so forth. The reports take only moments to create. The information can be very personalized, since categories that you have established will be used. The categories can be generic or can be as detailed as you want.
- Reconciling the account is almost fun. Before we got Quicken, keeping track of our personal finances used to be a chore my wife Jill dreaded. If there was a difference in the account it used to take a long time to locate the error. In fact, since it was such a chore, too often the reconciliation was never done. With Quicken, however, the reconciling steps are very quick and easy. Usually if there is a difference in the account, it is because of a handwritten check. This can be avoided by always using Quicken to write the

checks, but if this is not possible, it is still easy to go first to the handwritten checks to see if there are any errors.

Quicken's automatic reminder system helps to remind you of tasks such as:

- Checks to print.
- Electronic payments to transmit.
- Actions to take on investments.
- Recurring expenses for which you may not receive bills.

Use Quicken's automatic backup reminder to make sure you do not forget to do regular backups. It can be set to automatically remind you to do this always, weekly or monthly.

Quicken is easy to set up and use. The tutorial/assistant practically does all of the setup work for you. (See Figure 4.3.) If you have purged and organized your finance papers (bills, receipts, etc.) according to the PEP process described in Chapter 2, it is a simple process to convert to an electronic system.

1. With a few simple prompts you can choose the type of account (bank, credit card, investment account, etc.) and provide a starting balance.
2. Setting up the categories is just as easy. When setting up the account, Quicken will let you choose the type of categories you would like (personal/business); and then you can personalize them with matching categories created for your paper system. (See Figure 4.4.)
3. When a check is written or a deposit is made, it is assigned to a category. Some examples of income categories are: salary, dividends, interest, and bonus. Expense categories might include: mortgage, utilities, automobile, travel, and postage.

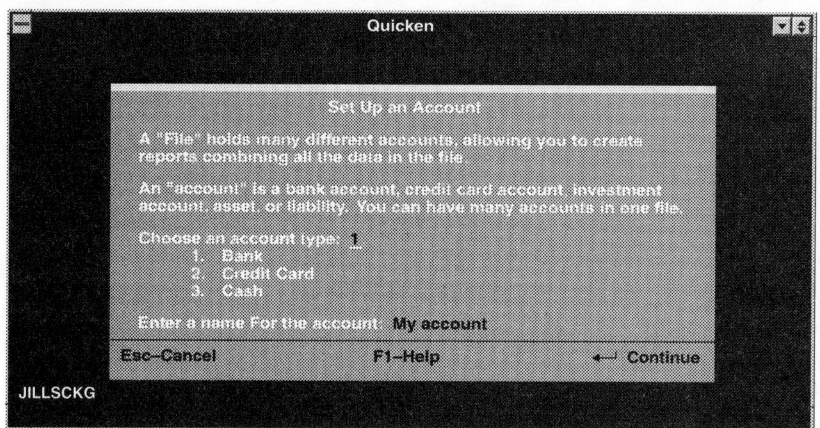

Figure 4.3 Setting up your Quicken account.

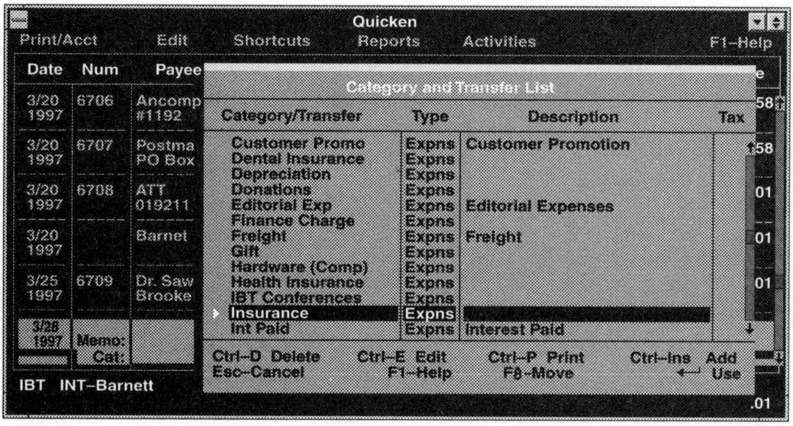

Figure 4.4 Quicken category listing.

4. Categories can also include subcategories; for example, utilities may include the subcategories of phone, electricity, and water.
5. Quicken is just as easy to use as it is to set up. When you have a check to write, click on "write/print checks" from the main menu, and presto, a blank check appears on the screen in front

of you. Type in the name of the payee and the dollar amount of the check, tabbing between fields. When you tab after entering the dollar amount, the system will automatically fill in the dollar amount in words; then you tab down and enter a category. If you are unsure of the categories you have set up, you can bring up the list of categories. You can scroll through the list, highlight the one you want, press enter, and it will be entered on the check screen (Figure 4.5).

6. Using the register (Figure 4.6) is also just like using the register in your old-fashioned checkbook, except that it is much easier. As you make an entry, Quicken's QuickFill feature recalls previous transactions as you type a few keystrokes. If the transaction that automatically appears is the one you want, just press enter; or you can keep on typing to recall a different transaction or add a new transaction.

7. Reconciling the account is very easy. All you do is open the reconcilement screen (Figure 4.7) and enter the ending balance and date of the statement (the beginning balance and date is already filled in from the previous reconcilement). Then you

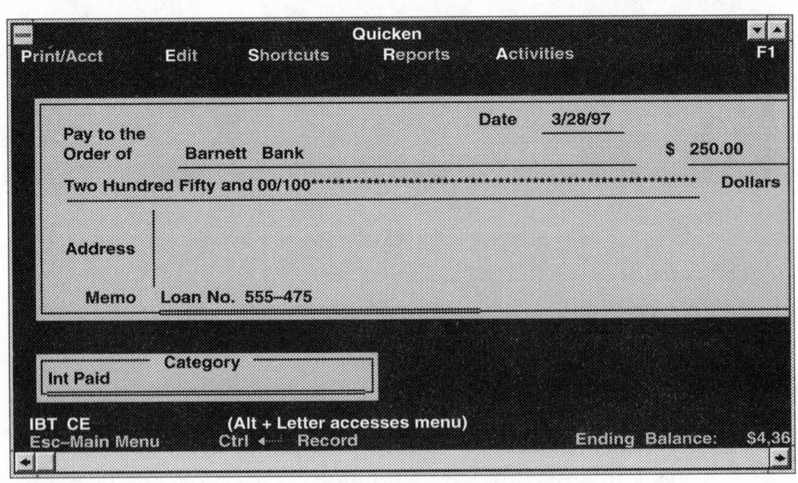

Figure 4.5 Quicken check screen.

Quicken

Print/Acct Edit Shortcuts Reports Activities F1–Help

Date	Num	Payee · Memo · Category	Payment		C	Deposit		Balance	
3/18 1997	6372 Memo: Cat:	Ford Credit PJ A970 812J Auto: Loan	553	54				6,521	17
3/18 1997	6373	Suntrust Bank South Florida 230081875001 Auto	222	60				6,298	57
3/18 1997	6374	First National/Chicago Heights 00008208325 50	67	00				6,231	00
3/18 1997	6375	F1. Prepaid College BG 1921185883 College	85	62				6,145	95
3/18 1997	6376	F1. Prepaid College QG 1921101634 College	44	93				6,101	02
3/18 1997	6377	F1. Prepaid College KG 1921186394 College	44	93				6,056	09

BARNETT (Alt + letter accesses menu)
Esc–Main Menu Ctrl ←┘ Record Ending Balance: $2,659.81

Figure 4.6 Quicken check register screen.

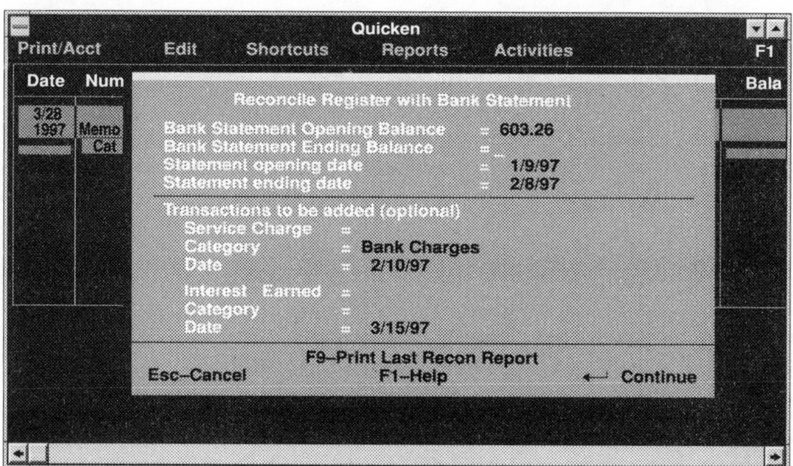

Figure 4.7 Quicken reconcilement screen.

just mark each cleared item. If you need to make an adjustment to your register, you can go back to the register, make your change, and then return to the reconcilement screen. After you have finished marking all of the cleared transactions, check the reconciliation summary to be sure that there is no difference,

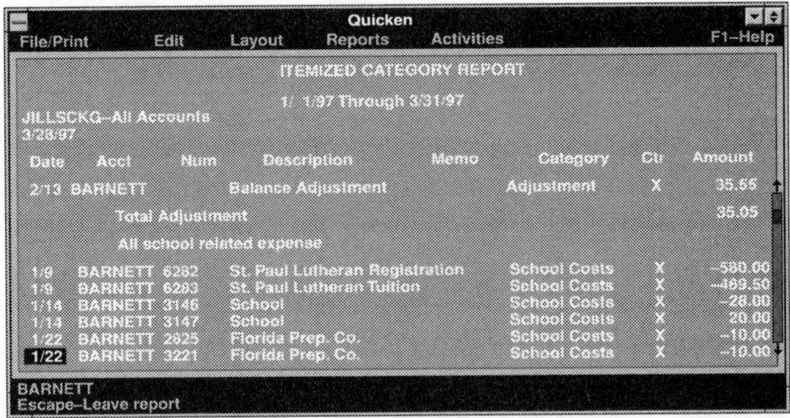

Figure 4.8 Quicken report.

and follow prompts to let Quicken know you are done. You can also print a reconciliation report if you wish.

8. The reports Quicken can produce are great when you want to view your finances from different perspectives. (See Figure 4.8.) You can create cash flow, net worth, and budget reports, and transaction, summary, and account balance reports. You can customize any of the reports by using a filter, and you can change the layout and display of your reports. Another helpful feature is the ability to display your finances using graphs. You can create line graphs, pie charts, and bar graphs using Quicken data.

ON-LINE BANKING

If you wish to move your money management to a whole new level of control, complement your newly set up electronic finance package with on-line banking (Figure 4.9). NationsBank is one of the banks engaged in utilizing the home-based technologies to increase customer services. The program is called NationsBank MYM (Manag-

Figure 4.9 On-line banking.

ing Your Money). NationsBank's service is typical of the services available with most larger banking institutions. Through a modem on-line banking enables the user to:

- Check up on the account(s) at any time as many times as desired. The user can bank 24 hours a day.
- Check balances and posted transactions in checking, savings, and credit card accounts.
- Transfer funds using a personal home computer. You can transfer funds from one account to another quickly and easily.
- Track money disbursement with almost no data entry.

- Pay your bills on-line (if you sign up for this separate service). You can schedule payments in advance and pay almost anyone.
- Get help with tax preparation. MYM automatically calculates the tax implications of your transactions on-screen.

NationsBank MYM, for example, is a complete, easy-to-use, personal financial manager that can be a time-saver even if only used for the basics (downloading transactions from the bank, reconciling accounts, and categorizing transactions). The program is in a compatible format with a variety of other money management programs. Much of the data recorded in these programs (e.g., Quicken—DOS versions 4–8 and Windows versions 1–5) can be imported into the banking program. The bank has also established a toll-free technical support system.

You spend only a minimal amount of time on-line (usually less than one minute). Most of the work is done either before making contact with the bank (selecting the desired actions or transactions, etc.) or after disconnecting. In addition, NationsBank seems to have used every precaution to protect the security of the program.

On-line banking complements the use of a money management software like Quicken. You will not only gain control over your finances, you will do it in much less time!

SUMMARY

It would be a shame if you had a computer and didn't take advantage of its most useful applications—controlling your time and money.

A good PIM used for organizing your tasks and planning out the accomplishment of your main objectives will prompt you to *act*. It is the consistency with which you work on the tasks related to your goals that ensures their successful accomplishment. A PIM reinforces that consistency.

Complement this with the adoption of an electronic finance application and you will quickly enjoy more control of your finances and a less stressful life.

FOLLOW-UP ACTION POINTS

✔ **Choose a personal information manager (PIM)** that best matches the nature of your work. If you use a companion product like a personal digital assistant (PDA), ensure the PIM you choose works with it.

✔ **Get familiar with your PIM** by studying the PIM's help section and through trial and error.

✔ **Convert from your existing paper-based system to the PIM.** Begin with the calendar section and add other vital information over time.

✔ **Keep your PIM open on the computer**, so it is easy to enter information, reminders, and so forth as they come up.

✔ If you haven't already done so, **get yourself a personal finance software package like Quicken.**

✔ Follow the Getting Started instructions of your finance package, and **get your money management converted to the new electronic system.**

✔ **Investigate your bank's on-line banking services.** Do what is necessary to **get up and running on them.** If your bank does not provide these services, find a bank that does and start using their services.

PREVIEW

The computer can be a wonderful asset when you're on the road. In this chapter you will learn:

☞ That using the laptop computer while on the road can dramatically improve personal productivity.

☞ That using electronic tools successfully begins with proper preparation before you even leave the office.

☞ What to look for and how to overcome difficulties connecting your electronic tools to your server or the Internet.

☞ Behaviors and methods of processing work that take advantage of your laptop as a productivity tool while on the road.

5

Road Warrior Wisdom

I purchased my first laptop computer in 1985. They weren't called laptops at the time. In fact, they quickly picked up the dubious name of "luggables" because they were so heavy. Mine was a Dataview 25. It had an Intel 8088 processor. It didn't have a hard drive. You had to carry and connect a second drive to the computer in order to use it. It also didn't have a great deal of functionality. There was no connecting to the Internet or to any network at the time. I used a software package called SmartSuite. It had a database application module, as well as a spreadsheet and word processing capability.

I vividly remember walking through the Schiphol Airport in Amsterdam carrying a 40-pound suitcase packed with enough clothes for a two-week trip along with 20 pounds of computer equipment in freezing rain (Figure 5.1). It was no party.

I tried my best to make use of this tool, which had cost me nearly four thousand dollars. I carried it around faithfully for several months until my arms could take it no longer. It just didn't make sense to bring along such a tool when it could only do a limited number of tasks, which in all honesty were more easily done with pen and paper.

It wasn't until some years later that I bought my next laptop computer. It was a 386 with 4 megabytes (MB) of random-access memory (RAM) and a 60 MB hard drive—state-of-the-art at the time. The hardware and software had improved dramatically during the interim. I was able to use my laptop not only to execute word processing and spreadsheet activities, but to connect to my home server and exchange information as well as organize my activities with a personal information manager (PIM). The computer itself weighed about eight pounds (a big difference from lugging around my Dataview 25). I have since graduated to my third-generation notebook computer. It has a Pentium processor, a color screen, 16 MB of RAM, a 500 MB hard drive and a 28,800 baud modem—not state-of-the-art at the time of this writing but certainly much better than previous models. Advances in processing power along with software breakthroughs have created many uses for my laptop computer.

You can expect continuous improvements with the technology (both hardware and software). In the not-too-distant future, I see videocon-

Figure 5.1 Lugging 60 pounds through the airport.

ferencing technology eliminating the need for much of the travel we are currently doing. We are not there yet, but that day is on the horizon.

Surely, technology will change how, why, and when we travel. Let's see how we can make our electronic tools work best when we're on the road.

USING A COMPUTER ON THE ROAD

Before I go into all the different uses for a computer when you're on the road, I want to say a few words about work behavior on the road.

In the course of doing this book, I've read every conceivable magazine and book on the topic of computerization and its impact on people's work behavior. I've interviewed a number of people who travel extensively on business. I have received numerous survey responses from seasoned road warriors through our corporate Web site on the Internet. I have had the input of many clients who have done PEP and use computers when traveling. One of the computer manufacturers' biggest selling points for notebook computers is the fact that you can use these machines while on the road and therefore get more work done.

It is not uncommon to find advertisements or articles written about people who claim that they can do lots of work in their hotel in the evenings because they have a computer. In the past, this would be work they would have had to do in the office. Seasoned travelers explain how they can use the time while waiting in airports and while on airplanes to catch up on numerous activities. I don't argue these points—they are true. But using a computer should not necessitate working longer hours; it should make you more productive. The amount of time spent working is often unrelated to how productive you are. In fact, experience teaches that a balanced life has more impact on productivity than the time devoted to so-called work.

There is a hard-nosed practical reason for my views. If you put limits on your work time, you'll be forced to figure out how to get the important things done within that time. Most of us have too much to do to begin with—more than can be done even if working 24 hours a day. If you leave your schedule open-ended, what will prompt you to get clever and resolve the workload and the quality of the work you are doing? Just because there are electronic tools that allow you to work virtually anywhere and anytime doesn't mean you should!

Okay, this may be a peculiar point of view. Nevertheless, we have many examples of people increasing their productivity threefold by employing *good work practices*. They normally accomplish this productivity improvement with less effort, and quite often without high-tech tools! If there is that much potential for productivity improvement, why not use it to buy more time—for yourself?

With that said, the computer certainly can be a productive tool for the road warrior. Here are some examples of what a road warrior might accomplish:

- For calendaring, whatever PIM you may be using on your desktop at the office can be used in your laptop.
- Letters can be prepared while on the road. If a letter needs to be typed based on a meeting or appointment held on the road, you can write it immediately while it's fresh in your mind.
- Any offers or proposals as a result of your trip on the road can be prepared immediately.
- Reports based upon the activities of your trip can be prepared on the spot.
- Because you have your PIM, you can adjust your planning as a result of your trip.
- Electronic notes can be taken during meetings.
- Memos can be written immediately.
- You can connect to your server to send and receive information.
- You can retrieve E-mail from your server and respond to it immediately.
- You can print documents if you have access to a printer.
- You can access information such as your address list or personal details such as account numbers when needed.
- With the proper equipment and a laptop you can make a multimedia presentation.
- Should you need to transfer any files, you are able to do so with your modem and laptop connected to your server.
- You can connect to the Internet or the World Wide Web to access information, receive and send E-mail (Figure 5.2), and transfer files around the world at a low cost.
- You can record your travel expenses with the proper software.
- You can work with and maintain any spreadsheets you may use.
- You can access customer records through a connection to your server and company information.

Figure 5.2 E-mail via hotel telephone.

These and many other functions can be executed with today's state-of-the-art computers.

WHAT TO CONSIDER WHEN TRAVELING

Proper preparation *before* you leave the office makes a world of difference when it comes to being able to utilize a computer and other electronic tools on the road.

Using a Travel Agent

I recommend that you find and use a good travel agent. While it may be possible to get a cheaper plane ticket by going directly to Internet travel sources, you have to weigh the time and effort (and thereby cost) of doing the search work yourself against delegating this function to a travel agent. Good travel agents can do much more than simply get the best fare.

- They have access to the latest changes in flight schedules.
- They have access to hotels providing the best in business services.
- They often negotiate the best deals for you.
- You have someone to complain to if your travel experience doesn't match what you expected from airlines, hotels, and travel services.
- Travel agents provide a vehicle to delegate to. (Bear in mind, those who are most successful in building a business are also those who best utilize the resources around them!)

However, there are times when it is easier to go on-line for travel information than it is to contact your travel agent. If you are traveling and you need to find out about a flight, try using Microsoft Expedia (http://www.expedia.com). Expedia is an on-line travel center that lets you book flights, hotels, and rental cars. It provides useful travel advice like how to book early and late flights to avoid crowds, choose alternative airports (e.g., Bromma Airport just outside Stockholm instead of Arlanda 40 kilometers away or La Guardia Airport outside of New York City instead of Kennedy Airport), travel on days other than Monday or Friday, and so forth.

For those of you who are bargain conscious, you can sometimes find inexpensive tickets on the Internet. Major airlines regularly offer Internet deals with savings of up to 90 percent. Check with the airlines you frequent to find out whether they offer E-mail notices for changing rates and how you can subscribe to their services.

Another excellent source of travel information on the Web is the Yahoo! travel directory (http://www.yahoo.com/recreation/travel).

Regardless of how you book your flight and hotel arrangements, plan it out so you have the resources you need to best keep in touch with your office and clients.

Planning Your Trip

What you bring on a business trip is determined by where you are traveling and what you hope to accomplish. Keep the following questions in mind:

- Where will you travel? Do they have different computer connecting conventions than the country you are traveling from?
- How frequently and from where will you be connecting to the home server or Internet?
- What will you be doing on the trip? (Presentations requiring equipment? Working while on a flight? Need to keep in telephone contact?)
- What will you need to produce with your computer while on the road? (Offers based upon the calls you make requiring a printed copy on-site?)
- Can you access necessary equipment locally, avoiding the need to bring it with you (for presentations, etc.)? Can you rent a mobile telephone that operates in the area where you are traveling?
- Do the hotels you will be staying at have a business center or facilities in the hotel room you can use to access E-mail, servers, printers, and so on that you may need?
- So far I have assumed your E-mail will arrive via an on-line service. If your E-mail arrives from an office LAN that is already set up with a modem node into which you can call, the process should be simple enough. Find out from your network administrator if any communications software and access codes are needed to get into the system.
- If you normally use a desktop computer at the office and travel with a notebook, you have to consider which computer

applications and files you will need to have with you. Often this means transferring information from one computer to the other.

Connection Issues

A good hotel should be able to deal with connection issues you may run into. Such issues might include:

- *Electric voltage differences:* The voltage might be 240 compared to 110 (normal in the United States). Most computers are built for both standards, so you will probably not need to concern yourself with this.
- *Internet connections:* The larger service providers like America Online and CompuServe will, in all likelihood, have local access numbers to connect you no matter where you travel. Smaller service providers may not have connection numbers outside of United States.
- *Modem connections:* Many countries have their own modem standards. Since yours may or may not function in a particular country, you should check beforehand to make sure you have a modem that will function where you are traveling. There are special modems that work in almost any country or can easily be configured to do so. (See TIP.)

If you travel outside the United States, you might consider a nifty modem that automatically configures itself to be compatible with most countries' standards. ClipperVom World PC Card Modem by Apex Data Corporation automatically configures the device for any of 30 countries in Asia and Europe as well as the United States. Choose the country, and the modem is seamlessly configured.

- Different countries may require adapters/plugs to access electricity and telephone lines. Inexpensive compact travel kits, with every conceivable connection tool, are available. I highly recommend that you take one with you if traveling abroad. One such pack is called Surelink Travel Connection Pack. It includes all telephone and power adapters you'll need for getting on-line anywhere.

- When traveling outside the United States, try to get local telephone numbers for help with access to the Internet or with software problems you may experience. If you are finding it difficult to connect your server you may be able to find local help and avoid an astronomical phone bill.
- A recent survey of frequent travelers identified lack of technical support as the biggest bug they run into. You should note down and bring the "help line" telephone numbers for the software/hardware vendors you use. You might find it useful to establish a procedure for getting tech support from your company's information technologies (IT) department while on the road.
- Be certain that the telephone line you are connecting your modem to is an analog line. The digital line is likely to blow out your modem (if a hotel or company uses a switchboard, the line is digital).
- Try renting a cell telephone from the hotel you will be staying at, or possibly, from your car rental firm. Consider using services that provide telephone rentals while on the road. In the United States, World Cell provides telephone rental services for domestic and international travelers.

Security

Preparations for your trip must include security. Your computer is a valuable and expensive tool. There are savvy thieves out there who would love to get their hands on it.

Insurance

It is wise to consider an insurance policy that will help replace any stolen equipment. One such company in the United States, Computer Insurance Agency (800-722-0385), insures notebook PCs.

Travel Case

Having the right traveling case can be another way of securing your computer. When you consider the basic electronic tools, accessories, papers, toiletries, clothing, and so on one normally brings on a trip, it is not surprising that we tend to underestimate how much case space we'll need. One computer case on the market that seems to fit the bill is called TravelPro. It's a carry-on case (with wheels) that has a hidden compartment for your PC. You roll the luggage onto the airplane like any other carry-on, only inside is your computer protected and concealed from any thieves.

According to travel case manufacturers, more than half of first-time buyers purchase a larger case within a year! Ensure that any travel case you purchase is big enough to store everything you'll be traveling with.

Backup

As expensive as a computer is, probably what is of more value to you is the data that's on the computer. Therefore you need to back up all your computer information before traveling. It's probably best to make a copy of your important information on disk to carry with you while on a trip. Keep it in a place separate from your notebook computer, in case the computer gets stolen. You can always replace your computer, and this way you'll have your most current information.

What to Bring on a Road Trip

Mike is a consultant with one of the Big Six accounting firms. He travels three to four days a week. He was an early proponent of the use of electronic tools on the road. As he describes it, he started out by purchasing and traveling with every conceivable accessory and peripheral. He has been steadily eliminating what he brings ever since! Many seasoned travelers will tell you the same thing. Most eventually realize how tiring it can be lugging around these tools even with the most lightweight laptop. This is especially true when they realize they don't even use much of what they bring. For example, I have never found that carrying a printer on a trip was of much use, or even an extra battery. Even though I own a rather compact adapter kit, I normally take out and carry only the adapters I'll use. I occasionally travel with multimedia tools. I say occasionally because along with everything else, lugging multimedia accessories can mean carrying upwards of 15 to 20 pounds.

If you're a multimedia user, consider purchasing a laptop computer that includes the features you'll need on the hardware itself. It is much easier than lugging this extra equipment around with you. Or, contact the companies and/or event sponsors you'll be seeing and ask if they can provide you with an LCD display or projector, or possibly a CD ROM setup, so that you don't have to bring your own. Most convention halls, hotels, and even companies have these tools for outside use.

If your hotel does not have access to a phone jack, consider an acoustic coupler, which is a device you can use instead of a modem to make a direct connection between a computer and a telephone.

When traveling internationally, your cordless telephone may not work in different countries. One solution: Rent a telephone in the country you go to. Europe operates on a common standard called GSM. As a result, if you rent a phone in one European country, you can use it in nearly all European countries. It's normally much easier than lugging your own and finding that you can't use it.

What you bring has to do with where you are traveling and what you will need to do while on the road. (See Figure 5.3.) I suggest you consider bringing the following:

- Telephone extension cord several feet long.
- An extra AC extension cord several feet long.
- Small power bar that also includes a surge protector and extension cord with multiple outlets.
- A three-prong to two-prong plug adapter.
- For international travelers, the appropriate electrical plugs and telephone jacks for the country (countries) where you are traveling.

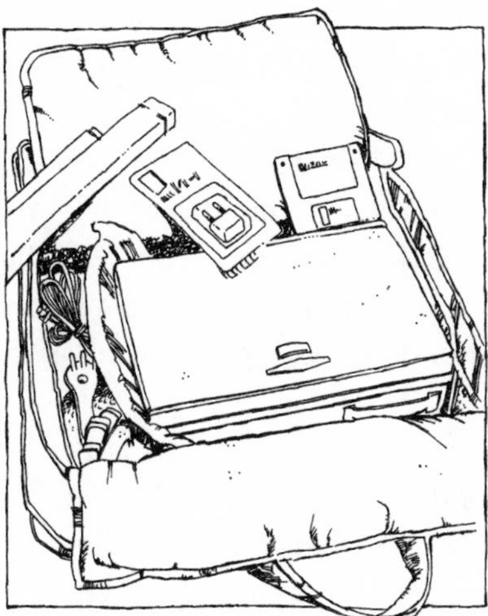

Figure 5.3 Packed computer case.

- Empty floppy diskettes to save work created on the road.
- Local access phone numbers of your on-line services.
- Rescue disk (bootable floppy) in the event your hard drive crashes on you!
- If traveling by automobile, a cigarette-lighter adapter that enables you to keep your laptop charged.

TRANSFERRING INFORMATION BETWEEN COMPUTERS

I avoid having to transfer information between computers because I happen to use my notebook both in the office and on the road. I am not alone in this. A Norwegian PEP participant has been traveling with different mobile computers for six years and has seen enormous productivity gains as a result, especially after the portables became so light. Ever since, the Norwegian has been using the same computer at work, while traveling, on-site with the customer, and at home.

Many of us, however, use a desktop in the office and a notebook while traveling. If that is the case, before you travel, transfer current versions of your files from your desktop to your notebook (assuming your files are not too big; i.e., do not exceed 1.44 MB of information—the size of current diskettes). If your files are bigger than 1.44 MB, the process becomes a bit more complex. You can use a software product like PKZIP to compress your files, but if you have big files and a lot of data to transfer, I wouldn't recommend it because you'll be using a lot of disks, which can be cumbersome to install on the notebook.

If your notebook and desktop computers are in close proximity, use a cable connection and a third-party software product like Laplink (by Traveling Software Corporation) to transfer files. Install the software in both computers and make the cable connection. With a few simple commands you can quickly transfer files between the computers. Laplink also enables you to do this procedure through modems while on the road.

Another option is to use an Iomega Zip drive (described in Chapter 3) that plugs into both computers. The Zip acts as the computer's hard drive. Files remain on the Zip and can connect to either computer.

Of course, if your desktop and notebook computers are part of a network connection, ask your network administrator to find the fastest way to update the files on your hard drive.

Whichever method you use, set up a simple and easy transfer system to use before your trip and upon your return.

ALTERNATIVES FOR STAYING CONNECTED

Personal Digital Assistants (PDAs), Palmtops, and Handheld PCs

PDAs have not been embraced by the market to the degree that manufacturers had hoped. Handheld organizers have been around many years. You will recognize such products as Newton, the Wizard, Omnigo, HP100/200, PSION, and many others. A PDA, or any of the other smaller handheld electronic organizer devices found on the market for that matter, are computers, albeit less powerful and with fewer functions than your notebook or laptop. PDAs have been described as a companion product to your desktop or notebook computer, as an information appliance providing you with basic data such as names, addresses, and telephone numbers, and more recently as a communications product enabling you to keep in touch with your office or the Internet. Earlier organizer or PDA products were not very powerful and had limitations that made them less than optimal. But that has changed and the products that you see on the market now and that will continue to be developed are rapidly gaining in popularity and usefulness.

With the advent of the Windows CE operating system for PDAs, you can expect to see many new products on the market with a lot more functionality.

Windows CE makes PDAs easier to use as well as more useful in several respects:

- Since the Windows CE operating system emulates Windows 95, users do not have to relearn a new operating system.
- Windows CE makes synchronization of information as easy as pressing a button. This is useful when it comes to maintaining a personal information manager on your desktop as well as in your personal digital assistant.
- Windows CE includes "pocket" versions of Word, Excel, and Internet Explorer. These pocket versions include most of the used features of the desktop versions of the software.
- Windows CE provides a PC card slot for modems. In some cases the modems are already installed in the hardware of the PDA. PC card modems for fax, cellular fax, and wireless, as well as paging cards are or will soon be available for these PDAs.
- Windows CE opens the door for many add-on products and applications. Some 300 application developers are producing products to enhance PDAs operating on Windows CE.

Whereas in the past PDAs have mainly served the purpose of being a personal organizer, with Windows CE and chip and software advances, these small PDA appliances become pocket versions of your laptop.

Depending on the model, PDAs in general can now:

- Be a full companion product with your notebook or desktop computer.
- Easily synchronize the information between your desktop and your PDA.
- Be a complete information appliance that can access files and information on your desktop.
- Act as your daily organizer (personal information manager). Many PDAs enable you to handwrite input, as well as use a keyboard, take notes, and many other things.
- Exchange files and synchronize data by both cable and infrared exchange.
- Access information such as contact names and numbers, your schedule, your calendar, and databases, and execute search functions for information.
- Become your communications vehicle. With a proper modem, a PDA can connect to your network at home and to your Internet service provider, whereby you can send and receive E-mail, faxes, and voice mail. Your PDA can also act as your pager with the proper hardware.

PDA/Palmtop Products and Their Usage

I am not going to attempt to review all the different PDAs and palmtops on the market. But for the sake of seeing how one might be able to utilize these products while on the road or otherwise, I have included a brief description of several of the more popular products on the market and the ways in which some people use these products to increase their productivity.

Pilot

One of the more popular PDAs on the market is the Pilot. I tested the Pilot and came to like it a lot. It is one of the smaller and more lightweight of the PDA products. The model that I tested had four functions: address book, scheduler, note taking, and to-do lists. You enter information with the stylus using a special kind of script called Graffiti. The most nifty feature of the Pilot is its automatic synchronization functionality. Connect the Pilot with your desktop (a special cord stored in the cradle comes with the Pilot to enable this) and press the hot sync button and the information on both computers is automatically updated with the most current data.

Even though I like the Pilot, I stopped using it. The main reason was the PIM in the Pilot was not the same one that I was using on my notebook and therefore I ended up having to enter information twice whenever I chose to carry around the Pilot instead of my laptop. Recently, however, Pilot has released a version that enables you to use many different types of PIMs other than their own proprietary one, as well as a communications function to send and receive E-mail. Pilot's drawback: You cannot transfer files from your desktop to the Pilot.

PSION

Lee is a PSION user from the United Kingdom. The PSION is one of the more sophisticated palmtops on the market. With the PSION, Lee is able to do word processing and spreadsheet activities and use an address book, a calendar (the calendar is unique in that priority tasks will show up in it if noted), external modems, associated software for wireless connection, as well as voice messaging and file synchronization. Lee's comment after beginning using the PSION: "I don't use a computer anymore." Downside: The PSION has its own operating system and you have to learn it from scratch. If you are a Mac user, there is no interaction with a Mac desktop.

Wizard

Jim works for a large chemical company in the Midwestern United States. His travel is limited, but when he does travel he brings along his Wizard. Jim uses it primarily for calendaring and address functions, as well as being a depository for miscellaneous information, such as Social Security numbers, credit card numbers, passwords, personal identification numbers (PINs), and anything else he might need to access in an emergency. He also uses it for all his financial management. The Wizard has many of the same functions as the PSION and Newton.

HP-100 and 200 Series

Ruth is a personnel director in a large bank in Europe. She finds using her HP-100 to be a great organizing tool. She uses it as a diary for recurring events and noting priority to-do's. It has telephone, address, and notes functions. The category function under notes is especially useful since it allows you to file different notes under several categories so you can task out projects. It is also possible to export notes taken on the HP into Word or Excel documents on the desktop. Carrying her HP-100 around in the corporate office enables Ruth to have access to the information she needs immediately, even if she's away from her desk.

Newton MessagePad

The Newton MessagePad is one of the original PDAs. The MessagePad 130's basic programs include the following:

- Note Pad—A convenient way to record information, draw maps, or create messages on the fly.
- Name Cards—A basic address book that allows the user to not only create a contacts database file but also use the MessagePad as a dialer to make calls.
- Calendar—A setup for recording scheduled activities, appointments, to-do's, etc., which will accommodate repeating appointments and activities and has an alarm that can be set to remind you about a specific activity.
- Pocket Quicken—finance software.
- Fax modem (a bit bulky to plug in), or PCMCIA fax modem card (expensive).
- E-world—Apple's on-line service for E-mail exchange with popular on-line services.

Deciding whether to use a PDA or palmtop depends on what you want to use it for. Nothing beats the Pilot if you intend to use it as a companion to your laptop—retrieving addresses, calendaring, or note taking. However, if you intend to use your PDA for communications, consider other products like the Newton or PSION. Many PDAs, with the proper modem, make it possible to fax, E-mail, and connect to your network.

PDA Connectivity Issues

Many PDAs allow for wireless connections. With the right cellular-ready modem, you can connect the PDA to telephone lines. A special radio modem called a packet radio allows you to send and receive wireless information to and from the ARDIS and RAM Mobile Data networks. Many PDAs also allow for paging cards. A paging card is the same as a typical pager you would find on your belt, except it's located in the PDA itself. The advantage of having a pager card in the PDA is that it can display longer messages than a traditional pager.

Using Your PDA Efficiently

There are several work behaviors that make the use of your PDA that much more productive.

- PDAs and palmtops have difficult entry systems, even if you're using a stylus. If possible, use your desktop to compose documents and drag and drop the documents from your desktop to your PDA.
- When you receive correspondence (whether E-mail, voice mail, or faxes) through your PDA, apply the same principles as you would with your computer. Process it immediately.
- Keep your PDA and desktop information synchronized. By doing so, in effect you're creating a backup of the information. It's always useful to have another source of backup.
- Talking about backups, your PDA information should be backed up as well. It's probably more important to do so with a PDA than with your desktop, since it's much easier to misplace a PDA than a desktop. The quick and painless way to back up your PDA is to get a PC card. Usually 1 MB or 2 MB static RAM cards can store information from your PDA.

PAGERS

Up until recently pagers were of two types: digital, which could only receive a phone number; and alphanumeric, which not only gave your phone number but provided you with a limited amount of text as part of the message. In both cases, you needed to find a telephone and return the call.

Two-way pagers have now become the rage. Two-way pagers allow you to receive text and numbers as in the alphanumeric type as well as enable you to respond to that text. Not only can you receive text but you can receive and send E-mail as well as pick up voice mail with these two-way pager products. Skytel is the service provider for these two-way pagers, and they have released one of the smallest pagers on the market, called Skywriter.

Mike, a facilities manager with one of the larger accounting firms swears by Skytel services. While traveling, he gets nationwide coverage and Internet access, including the ability to communicate with Internet E-mail users through the pager. His pager notifies him of voice mail through a toll free voice mail system to which he can reply directly from his pager.

Motorola has released its own two-way pager called PageWriter. PageWriter comes with a 46-key QUERTY keyboard, which makes composition easier. PageWriter also enables users to silence any incoming messages, other than those you've specified.

I'm sure we'll be seeing many more manufacturers releasing two-way pagers in the near future.

CELLULAR AND DIGITAL PHONES

Arguably, one of the most useful tools on the road is the cell phone. Of course, as is the case with a pager, coverage is an issue with cell phones. This is not much of a problem when traveling domestically, but traveling abroad can cause difficulties. Telephone standards in other countries make it nearly impossible to use just one telephone when traveling internationally.

Up until very recently cellular telephones were limited in their features. Size, battery length, and design were the main differentiating factors for the purchase of a cell phone. However, technology has improved and there are now mobile phones on the market that are referred to as "smart phones."

SMART PHONES

A smart phone not only functions as a telephone, but can also exchange fax and voice mail as well as data information (with a built-in cellular modem). With the right software, such as Unwired Planets' UP.Link, you are able to set pointer and reference settings to program the phone to retrieve E-mail and even connect to private networks and the Internet.

Take it a step further and you have Nokia's telephone products. In Europe it is referred to as Nokia 9000. In the United States, PCS 1900. It's not only a telephone, but can also relay and receive faxes as well as access the Internet and send and receive short messages and E-mail. It even has an LCD display and keyboard with a PIM application built into the telephone. With the Nokia 9000 you could throw away your PDA and pager and have only one machine, a telephone, enabling you to do many of the things that you would do otherwise with the computer.

Other manufacturers have released their own smart phones including Mitsubishi's Wireless Communicator Mobile Access and AT&T's Pocket Net Phone.

Telephone etiquette: Avoid taking and making cell calls while dining or having a meeting. If you must call, excuse yourself from the table to do so.

WIRELESS SERVICES

A great way to avoid the hassle of connecting the telephone by wire to retrieve messages is to use a wireless service. Here are a number of wireless service providers.

WyndMail, from Wynd Communications Corporation of San Luis Obispo, California. WyndMail is a two-way messaging service. With WyndMail you can receive and send E-mail and voice mail through a wireless modem. When you receive E-mail, WyndMail sends a wireless signal to your computer, beginning with a "snapshot" of the full text. The snapshot includes information on the sender, message subject, and the time it was sent. You can decide to receive the message, store it on WyndMail's Internet server, or delete it.

ZAP-it. ZAP-it provides almost identical services to WyndMail but uses the existing RAM Mobile Data network, Motorola's two-way wireless data communications network that offers U.S. nationwide coverage in major metropolitan areas (7700 cities), to transmit data via radio waves. Messages are then delivered to the ZAP-it messaging computer and routed to their final destination. Downside: ZAP-it only functions in the United States and, unlike the telephone network, the wireless radio infrastructure is limited to sending short bursts of data, called packets. The more packets you send the more expensive the message.

Personal Messenger 100C Wireless Modem Card, by Motorola, based in Schaumburg, Illinois. Wireless modems can be a lifesaver, depending on the nature of your work. And PMCIA is designed to be as wireless as possible. Just slide it into any PMCIA slot and it is ready to go without the need to attach cords or cellular phones. You can send and receive messages and even tap into your network at the office and use any application you'd normally use. Downside: It will operate only through telephone companies or special service providers using CDPD (Cellular Digital Packet Data). This is broadly available in the United States but unavailable outside the country.

PageSoft Pro, a Windows software program by Socket Communications Inc. PageSoft Pro allows you to send and receive messages from desktop and notebook computers to alphanumeric pagers, the Socket PageCard data pager, or handheld computers. It allows you to download messages from the PageCard data pager directly into the software's in box. Remote E-mail users view their paged messages in the in box of the most popular E-mail systems (CC:Mail, Microsoft Mail, Newton MessagePad, or HP-100/200 LX palmtop computers). Downside: You need a pager and you are limited by the service's coverage.

Socket Wireless Messaging Services (SWiMS), administered by the National Dispatch Center. This network turns your pager into a wireless link to telephone, voice mail, fax, E-mail, modem, and pager. You are issued a personal 800 number. Messages and data are sent to your 800 number over conventional phone lines and messages are routed wireless to your pager. You can call SWiMS and play back your voice mail and forward your faxes. An international access number lets you retrieve your messages and forward faxes when you are traveling abroad.

Email Reader, from Millennia Software. Even if you don't bring a computer, you can still collect your E-mail from your desktop at the office using Email Reader. If your desktop is a Pentium PC running Windows 95 with a TAPI-compliant modem (available from U.S. Robotics, Diamond Multimedia, and Creative Labs), you can retrieve E-mail over an ordinary telephone line. Just dial your computer's number and speak. The computer talks back and reads your E-mail aloud! After you call your computer you can move through your E-mail messages via spoken intuitive commands, including "Retrieve new mail," "Previous messages," "First message," "Play message," and so forth. Ask "What was that?" and the computer repeats the last spoken word. Downside: You cannot reply to the E-mail in the same way. Nevertheless, a wonderful tool to be able to easily retrieve E-mail without having to carry any electronic device at all!

Wildfire. No matter how prepared and organized you might be when you travel, gathering information can still be complicated. Aside from having to adjust to all the physical needs (adapters, voltage, plugs, etc.) you may also find it difficult to connect all your different sources of information (voice mail, phone mail, faxes, E-mail, etc.). Services exist that take much of the pain out of sending and receiving messages (voice, fax, and E-mail) while on the road. One such service is called Wildfire, which uses voice recognition to converse with you. Wildfire, as Jeffery Kagan puts it in *Mobile Entrepreneur* magazine, is an electronic voice messaging system designed for businesspeople who must remain connected to the office even while traveling around the world. It answers your phone, takes messages, responds to voice commands, and lets you leave automatic messages for certain callers. It recognizes the voices of those who have called before (greeting them with, "Oh, hi") and if a call comes in while you're listening to messages, Wildfire will whisper in your ear to ask if you want to have the call put through or take a message. Wildfire anticipates future electronic office systems by marrying several technologies—telephone, database, PC contact manager, and voice recognition—into a seamless whole. Because Wildfire records the phone number of each call that comes in, you can respond to messages instantly by simply saying, "Give them a call," and Wildfire places the call for you. Faxes and E-mail messages can be forwarded to you in hard copy anywhere in the world. Wildfire represents the next generation of similar services available from some phone service providers. These services do not require a PC link and can be a cost-effective way to make sure you never miss a call.

WorldLink. WorldLink has similar features to Wildfire and acts as a universal in box for road warriors. Each subscriber gets a personal 800 number that receives and stores messages, whether in the form of voice mail, faxes, or E-mail sent via CompuServe. The system automatically pages subscribers when urgent messages arrive. It allows people sending faxes to annotate their messages with a voice message identifying the source and the subject. This way, subscribers can check their WorldLink account, hear who has sent them faxes, and choose what to print. The system allows subscribers to respond to E-mail over the telephone by choosing from a set of preselected options. Added plus: the service is inexpensive—only 25 cents per minute basic usage charge, with no setup or monthly minimum fees. (See Figure 5.4.)

Telephone companies such as AT&T and MCI in the United States, as well as larger non-U.S. telephone companies like Telstra in Aus-

Figure 5.4 Wildfire and WorldLink handle many aspects of staying connected.

tralia provide similar services, usually for a monthly fee. You would need to contact these service providers, explain your work process and travel destinations, and get a clear idea of each provider's service features to find the best choice for you. With so many services and products on the market, you are certain to find something that meets your needs and wishes.

FEELING CONNECTED

Many of us have become frustrated with voice mail systems. And if you are on the road and do not promptly reply to messages people get even more upset. To avoid this, leave a complete announcement message on your voice mail system explaining your absence and letting callers know that you will be checking in daily and will respond to any messages.

Your announcement message might sound something like this: "Hello, my name is Kerry Gleeson. Thank you for calling. I am not in the office at the moment as I am on an international trip and will be gone until the tenth of the new month. I will be checking my messages daily and if you will kindly leave a complete message including the purpose of your call and a good time to get back to you, I will do my best to respond quickly."

If you get connected to someone's voice mail and you do not want to listen to the entire outgoing announcement message, try pressing star (*) or pound (#), and you may be able to skip it and get directly to the recording beep.

HOW TO KEEP UP IN THE OFFICE WHEN YOU'RE NOT THERE

A colleague who works in your office should inform you of everything that has come into the office that day, and help you get things activated and keep things moving. This may be as simple as faxing copies of everything to your hotel so you can respond to them at some point during the day.

Most travelers get updated to some degree or another. What they do *not* usually do is:

- Check *all* messages/faxes/mail instead of focusing on only what is considered urgent or important.
- Process correspondence "now."

When you're on the road and don't process the work that accumulates in the office, you return from your trip with both the work you did while you were away and the work that's piled up at the office. A backlog like this is often difficult to dig yourself out of.

The best solution is to get the vast majority—80 to 90 percent of what comes in—relayed to you (either verbally or electronically). Once received, it can be processed or activated in some way immediately. A little bit every day goes a long way toward keeping on top of what you need to do.

ROAD WISDOM

Here are a few rules of thumb to make your trip more productive and so you don't come back to mounds of work:

1. Check in daily and process everything you can. I make it a point of calling the office two or three times a day from wherever I am. (If I am on the other side of the world I might cut it down to once. But I do call every day to make sure that I stay in touch.)
2. Process all electronic mail and faxes that may be forwarded to you that same day. Don't just look at the E-mail and decide that it is something that you can do later. Get clever about how you can get this completed and/or activated so it can be completed in the quickest time possible.
3. Process all activities that may have occurred that day while on the road. Do not allow things to accumulate until you get back. If you have an offer to make or proposal to write, type it up then and there. If you have expenses to account for, get them accounted for immediately. If you have to send E-mail messages or delegate something, get it done now. Schedule the time, while on the road, to get this done.
4. Delegate liberally.

I am not suggesting that you work at night to get this done. Get smart and organize yourself and get the work done within your schedule. It is possible! (See Figure 5.5.)

Figure 5.5 Earn yourself some lazy time by getting your work done during work hours.

SUMMARY

The road warrior does not lack for productivity tools. Many electronic tools truly deliver what they promise—get more done in less time. Speed is probably the most critical element to success in this day and age. PCs and PDAs dramatically increase speed of communication and the flow of information. Take care in determining which tools and services best suit your travel needs. Prepare for your trips. Process work daily and avoid backlogs. As difficult as all this may be, it is much easier than coming home to mounds of work!

FOLLOW-UP ACTION POINTS

✔ **Evaluate your travel needs.** How often are you on the road? Where do you normally travel? How closely do you need to stay in touch?

✔ **Decide on the tools and services that will best support your travel.** Do you need the full power of a notebook PC or will a PDA or handheld computer do? Can a PDA be a companion to your PC? If traveling internationally, would it be helpful to have a modem for use in most countries? Or can you abandon these tools in favor of a service like Wildfire? Make up your mind and get the tools you need.

✔ **Prepare carefully for your trips:**

a. Do a full backup.

b. Decide what supplies to bring and pack them.

c. Arrange to stay in hotels that make it easy for you to connect and collect.

✔ **While on the road, connect and collect your daily work from the office.** Process this work immediately. Do not let the work backlog!

PREVIEW

The Internet can be a wonderful resource to enhance both your personal and professional life. It can also be a waste of time. In this chapter you will learn:

- ☞ How to access the Internet and World Wide Web.
- ☞ How to organize Internet documents and E-mail messages.
- ☞ How to efficiently search for information on the Net.
- ☞ How to avoid information overload.

6
The Internet and World Wide Web

The Internet. What a resource! What a mess! Seems like everywhere you look, someone is touting the Internet and how it is changing the world as we view it. What is the Internet? What is the World Wide Web? (See Figure 6.1.) What is all the excitement? Does the Net live up to all of the hype? I am not sure the Internet is all that it's cracked up to be. One thing is certain: If you are in business, you need to know something about the Net, if for no other reason than not to appear too out of it! Apart from that, when you familiarize yourself with the Net you might even find it to be a useful resource.

The Internet is essentially a network of computers connected with each other. Much of the information on these computers is available to those who wish to access it. The story goes that the Internet was designed by the government in the late 1960s as a way to communicate in the event of a nuclear attack. This is just a popular myth. The actual story is contained in the book, *Where Wizards Stay Up Late: The Origins of the Internet*, by Katie Hafner and Matthew Lyon, published by Simon & Schuster. The book starts with a prologue dated September 1994, where several of the scientists and engineers who invented and created the Internet met in Boston to celebrate the 25th anniversary of the Net. A quote from page 10: "Bob Taylor, the Director of a corporate research facility in Silicon Valley, had come to the party for old times' sake, but he was also on a personal mission to correct an inaccuracy of long standing. Rumors had persisted for years that the ARPANET had been built to protect national security in the face of a nuclear attack. It was a myth that had gone unchallenged long enough to become widely accepted as fact. Taylor had been the young director of the office within the Defense Department's Advanced Research Projects Agency overseeing computer research, and he was the one who had started the ARPANET. The project had embodied the most peaceful intentions—to link computers at scientific laboratories across the country so that researchers might share computer resources. Taylor knew the ARPANET and its progeny, the Internet, had nothing to do with supporting or surviving war—never did."

Over the years more and more computers were added to the Inter-

Figure 6.1 The World Wide Web.

net, providing much more information. During the early period of its existence, the Internet remained in the domain of the government and universities. Now that has changed and there are many millions of people from around the world who are connected to the Net.

Basically, you can thank the World Wide Web for the Internet's exploding popularity. A few years ago, the World Wide Web (WWW) came on the scene—a collection of multimedia documents connected by devices called hyperlinks. Multimedia documents include not only words and numbers but also pictures, sounds, animation, video, and anything else that you can store on the computer. Until the advent of the World Wide Web, the Internet only used simple text and was rather boring and difficult to work with. Hyperlinks within World Wide Web documents have made accessing other documents as easy as clicking your mouse on the hyperlink, which automatically takes you to the linked Web site.

You can collect almost any conceivable bit of information through the Internet and WWW—anything from government census data to statistics on the team players of the Chicago Bulls.

But the Internet can also be a terrible waste of time. Let's see how we can take advantage of the Internet as a resource and avoid the time-wasting traps.

DEFINITIONS

Before proceeding, take a few minutes and familarize yourself with common Internet terms in the Glossary at the back of the book. Terms such as hyperlinks, TCP/IP, browser, WWW, and so on all have their own meaning within the Internet world.

GETTING SET UP ON THE INTERNET

To access the Internet you need to have four things:

1. The hardware—usually a computer and modem.
2. Communications software—to be able to use the modem.
3. An Internet service provider (ISP)—to connect to the Internet.
4. Browser software—to manipulate around the Internet and view WWW documents.

Hardware and Communications Software

If you have gotten this far in the book, I can only assume that you have a computer. Hopefully, you also have a modem that will allow you to connect to the Internet at high speed and communications software allowing you to use the modem. How quickly you can access Internet sites and download information from those sites depends on how fast your modem is. You'll soon discover trying to download video or graphics from the WWW can be a very time-consuming process if you have modem that is 9600- or 14400- or even 28800-baud. Should you have access to what is called the ISDN (Integrated Services Digital Network) line, you'll be able to download information much more quickly than through normal telephone lines. ISDN is designed for speedy transmission of digital information. Most ISDN cards are rather tricky to install, so it is best to let an expert handle the installation.

Internet Service Provider (ISP)

Next, you'll need to sign up with an Internet service provider. America Online (AOL) and CompuServe are two of the better known service providers. These are not the only service providers you can consider, though I suggest that you stick with one of these bigger on-line services. The larger ones will provide easier access to content not available with your smaller providers, and if you are an international traveler, you will be able to get local access connection through CompuServe and AOL practically anywhere in the world.

Browser Software

You'll also need something called Web browser software, which allows you to access E-mail and view WWW pages on the Internet. AOL and CompuServe both provide their own browser software when you sign up for their services. However, the most popular browser software packages are Netscape and Microsoft Internet Explorer. Combined, they account for about 80 percent of the browser software market. My recommendation for your choice is Netscape. Netscape has the largest browser market share. If a WWW page has been designed to be viewed through a specific browser, using another browser may give it a different look, such as pages (text and graphics) out of alignment and not representing what the site creator intended.

ORGANIZING AND UTILIZING BROWSER SOFTWARE

I am not going to attempt to write a text on the ins and outs of the different browsers available on the market. Instead I will cover, in general terms, how to compose E-mail messages, fill out forms, and transfer files through the Internet. Later in the chapter I will cover efficient means of searching for information.

Internet Connection

Assuming you have installed your browser software (or it was installed for you through AOL or another ISP) and depending on your browser, connecting to the WWW normally begins with double-clicking on the browser icon. In the case of Netscape it begins by double-clicking the Dial Up Networking shortcut icon. A dialogue box appears, calling for your password. Type it in and click on Connect and it connects you to your service provider. Double-click the Netscape Navigator icon and your browser home page arrives on the computer desktop. (See Figure 6.2.)

In the case of Microsoft Explorer, double-click on the Internet Explorer icon to both connect to your service provider and bring your browser up on your screen.

Like any software program, browsers will be able to execute a number of functions, and the best way to learn how to use this software is to go through its tutorial and help section.

BROWSER AND E-MAIL

Arguably the most useful function of the Internet is E-mail. Browser software makes it easy to compose, send, and receive E-mail. In the case of Netscape Navigator, once you're connected to a service

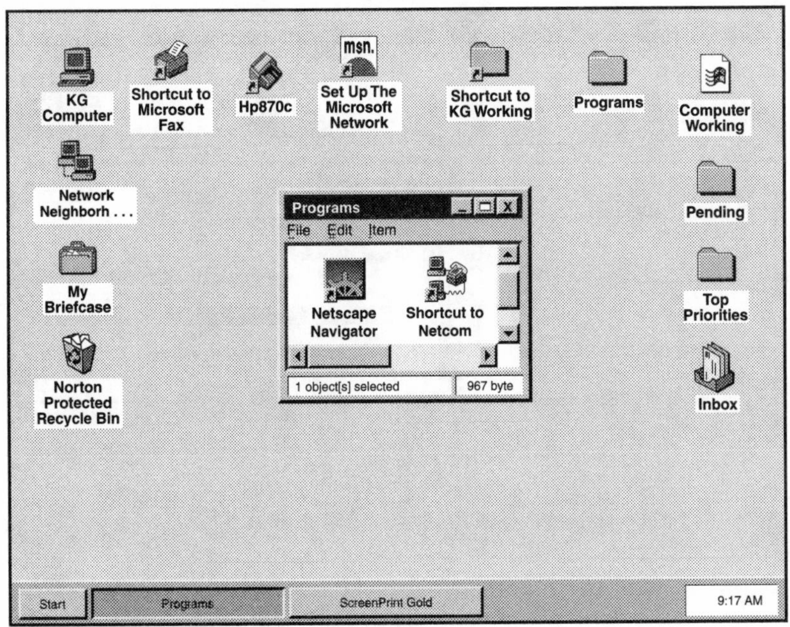

Figure 6.2 Windows 95 work space with Netscape icon.

provider, select Netscape Mail and your mailbox comes to view on the computer work space. (See Figure 6.3.)

Your ISP will post any E-mail that has been sent to you in your in box (the right side of the screen).

Organizing E-mail Messages

My guess is the vast majority of your E-mail will be deleted upon receipt and processing. But any E-mail you keep will need to be filed. Organizing your E-mail is easy. Browsers allow you to assign a category (folder) to the E-mail message, so you can access saved E-mail messages. In most cases, to create a category, select the E-mail message in your in box, click on New Folder, and type in the category name. To file your E-mail in an existing folder, simply drag and drop the E-mail message into it. (See Figure 6.4.)

For more information on organizing E-mail messages see Chapter 3.

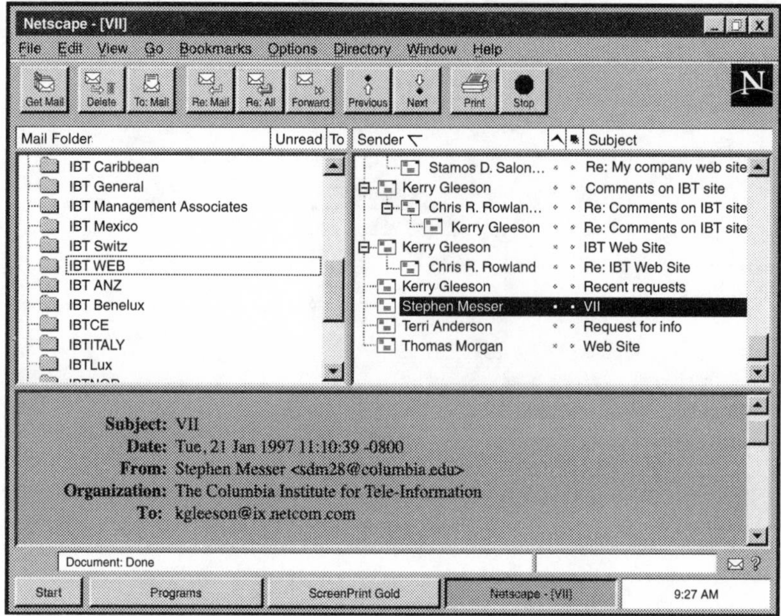

Figure 6.3 Example of Netscape mailbox.

Composing a New E-mail Message

To compose a new E-mail message, almost all browser software will direct you to an icon for New Mail Message and it will bring up an E-mail dialogue box to fill in. Type in the person's E-mail address, fill in the Subject field, and type in the text of your message. If you have the recipient's E-mail address in your browser address book, choose the appropriate address icon and locate the person in your address list. Follow the browser software instructions to select the person's address, and it will show up in the Mail To field. When you've finished composing the message (assuming you are still on-line), just click on Send to "mail" it.

Composing an E-mail Response

Most browser software allows you to choose to reply to a message and will bring up an E-mail dialogue box with the sender's E-mail address already filled in. One would normally use the tab key to

Figure 6.4 Example of folder tree.

move from field to field in the E-mail form. Type in your response, choose Send, and that's all there is to it.

Creating and Organizing an E-mail Address Book

Browsers can also memorize E-mail addresses and put them in an address book. Most browsers organize the addresses in alphabetical order, but it is possible to create general categories (groupings) within the address book list and file addresses under these general groupings. For example, I organize my address book according to those that work in my business and under the grouping/category Personal, making it easier to find the address I'm looking for. (See Figure 6.5.)

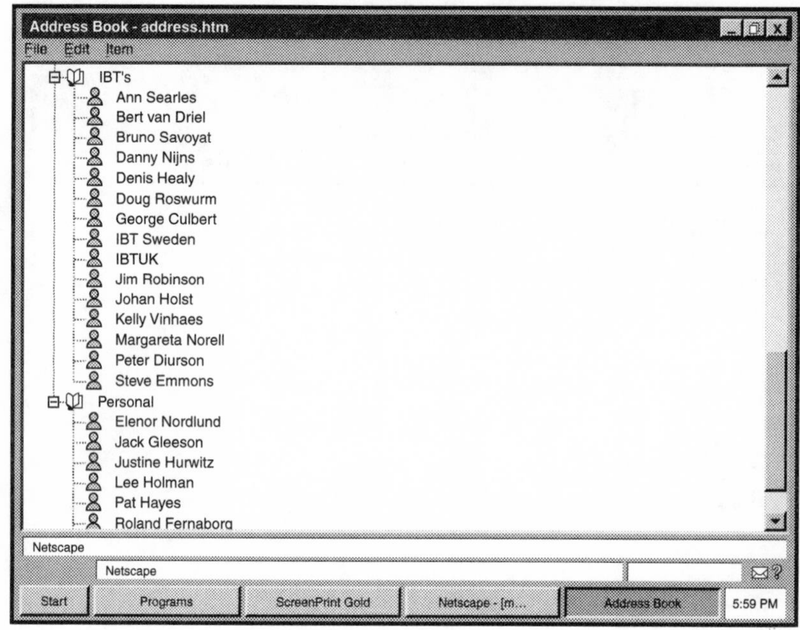

Figure 6.5 Address book with example of address list groupings.

Filling Out Forms

Filling out forms you may encounter on the Internet or WWW is much like working with dialogue boxes in E-mail composition on your browser. Forms on the Internet might be used to order products, utilize search services, or execute any number of other functions. Open the form and use your tab key to maneuver around the different fields. Mark the check boxes or option buttons, fill in the text fields, and follow the command buttons of the form. (See Figure 6.6.)

Attaching Files to an E-mail Message

The Internet makes it possible to transfer files from your computer to a distant computer or vice versa. Browser software makes it easy to attach a file to an E-mail message and send it off for viewing at the other end. While all browsers are not the same, each will have a way

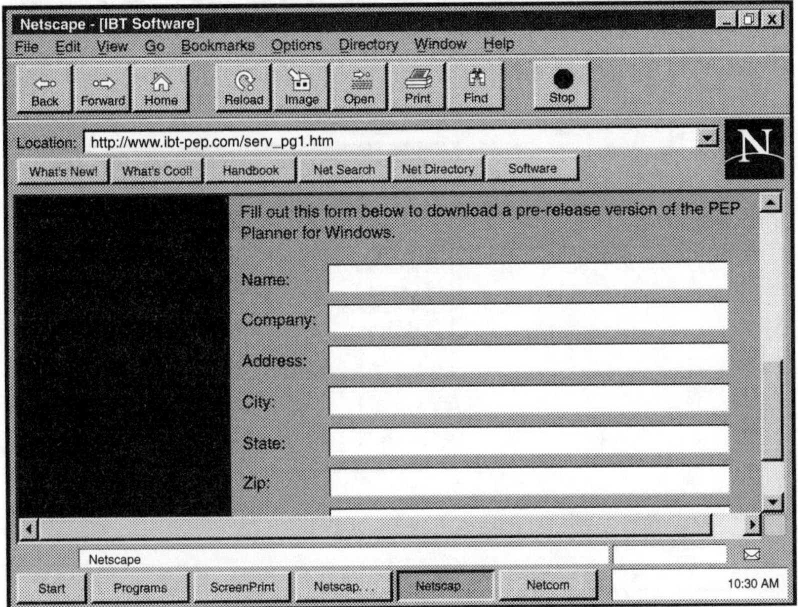

Figure 6.6 Example of World Wide Web form.

to attach a file during E-mail composition. In the case of Netscape, you have the choice of clicking on the Attachment field, clicking on the Attach icon, or choosing File and Attach File. (See Figure 6.7.)

Viewing File Attachments

While it may be easy enough to send and receive files, *reading* a file attachment can be difficult because many files can be viewed only through the application that created them. You may not be able to read a file if you don't happen to have the application that the file was created in. It's obviously unreasonable to install every conceivable application on your hard drive simply to be able to view a file you may or may not receive. Larger companies have software standards, so normally file transfer among colleagues within the company is not a problem. But you cannot depend on the same software standards from those outside the company.

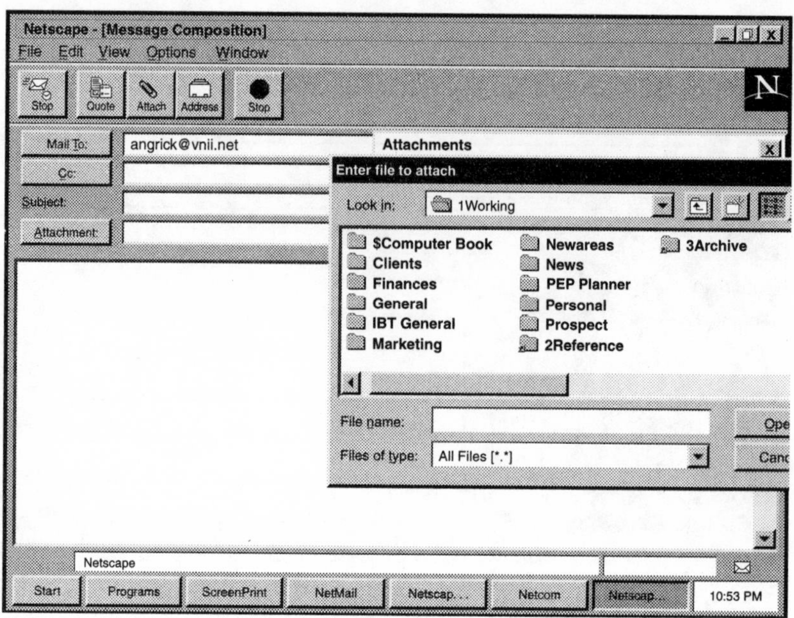

Figure 6.7 Example of attachment box.

A simple and inexpensive solution is to purchase a software package called QuickView Plus. With QuickView Plus you're able to view files created in 200 different applications (nearly every major software application on the market). QuickView Plus is an excellent product, and I highly recommend it to those of you who will be sending and receiving files through the Internet.

COPING WITH E-MAIL OVERLOAD

The amount of electronic information we all receive can be overwhelming. It can blind us to what we should be focused on. In this information age it is as important to know what information and communication you *don't* need as it is to know what information and communication you *do* need. Don't allow low-value information and tasks to enter your system. You need to aggressively screen all sources of information if you expect to have any chance of avoiding being swamped. Figure 6.8 illustrates how to filter communications and information, so you don't get unwanted materials in the first place.

E-mail Extravaganza

By creating a structure for your electronic documents, you have taken a big step toward controlling the flow of electronic information. The primary source of electronic information for most people comes from the Internet or LAN in the form of E-mail. E-mail is a marvelous tool. Most benefits from E-mail are obvious:

- Communication is simplified.
- E-mail is less costly than traditional mail.
- Loss of the message (information) is less likely.
- With a single keystroke, the message can be sent to many people.
- Remote access is possible.
- Receipt of messages can be verified.
- The documents are easier to organize.
- The documents are easier to manipulate, reorganize, and edit.
- It is easier to transfer documents.
- E-mail is speedier than most other forms of communication.
- With E-mail, you can append all previous communications so a complete history moves with the note.
- Some E-mail systems allow you to send yourself a message at a later date, another type of reminder system.
- E-mail makes possible the preprogramming of function keys to set up timesaving keystrokes for tickler system entries (calendar and appointment reminders), delete procedures, and the like.

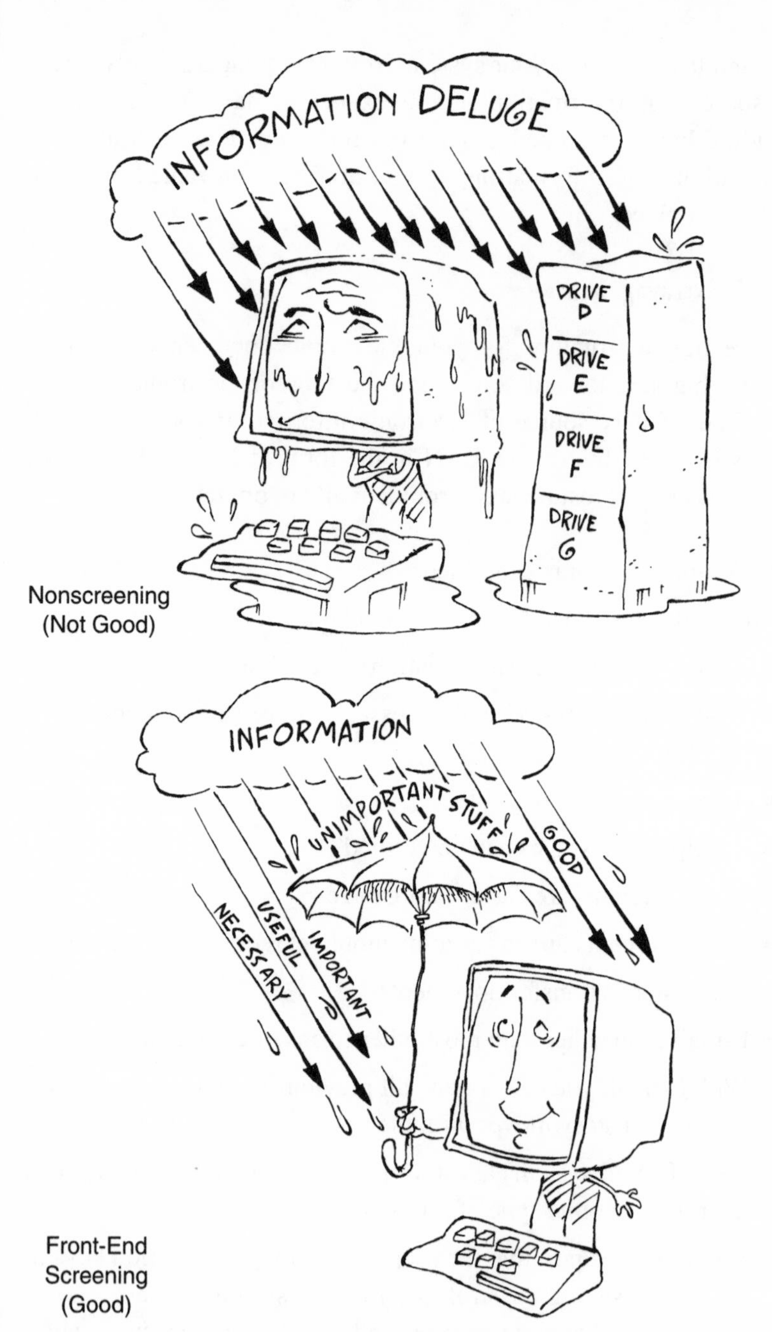

Figure 6.8 Screening information.

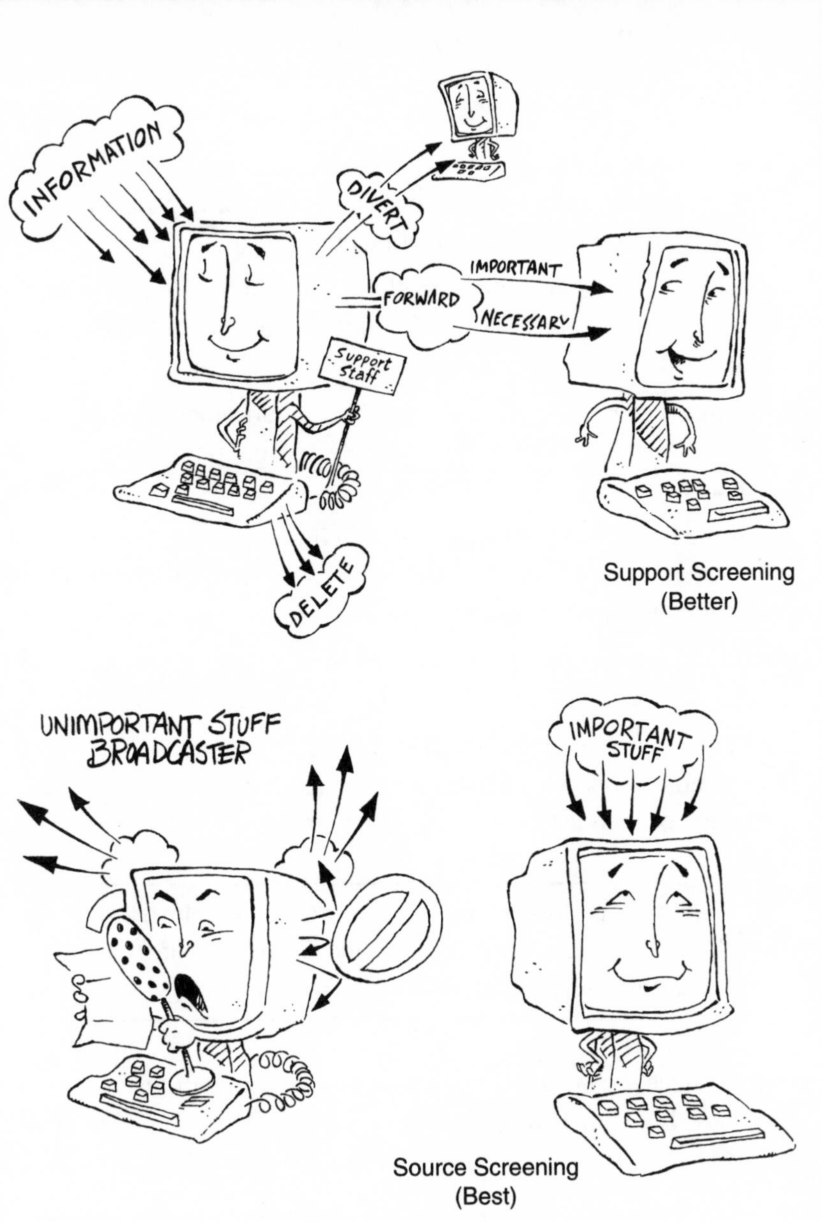

Support Screening (Better)

Source Screening (Best)

Figure 6.8 Continued.

Along with the good comes the bad:

- Because E-mail makes it easier to communicate, there tends to be more of it. Because there can be so much of it, people tend to ignore messages they deem unimportant.
- E-mail systems are subject to abuse. It is not uncommon to find people using it for things like advertising the birth of kittens they are eager to get adopted.
- E-mail can become the preferred form of documentation, as one of my banking clients put it, "to CYA (cover your [certain anatomy])."
- Some E-mail systems have irritating limitations like having to enter an E-mail document and follow a series of instructions to simply delete the message.
- Employees may not have switched from a paper to an electronic mind-set and may end up printing every E-mail message.
- The system can get overloaded. Typically information technology departments limit the number of messages allowed in any one workstation.

E-mail is a great tool to prep for a meeting. Instead of using meeting time for announcements, prepare and send the announcements beforehand as E-mail.

On balance, I consider E-mail to be an excellent tool for improved efficiency. But E-mail will overwhelm you if you are sloppy in dealing with it.

If you find yourself with lots of "pending" E-mail (messages you are unable to bring to closure because you are awaiting information, etc.), you can organize these pending E-mail messages by creating alphabetized folders—A-Pending, B-Pending, C-Pending, and so forth—and copy the message under the appropriate letter. You will find it more easily.

Handling E-mail the Right Way!

It is vital to apply the **Do It Now** principle when processing E-mail (as well as paper mail, faxes, and voice mail messages). Many of the problems you experience coping with the volumes of traffic you receive will resolve by simply ***doing it now***. But no matter how you cut it, if you have allowed yourself to receive it, you must handle it. The question you have to ask yourself is, Should you be getting it in the first place? **Sean Savage of Knight-Ridder Newspapers** covered solutions to this problem in a recently published article. If you have an E-mail address on the Internet and have an Internet provider (America Online, CompuServe, Prodigy, etc.), chances are you are receiving electronic junk mail. It is common practice for Internet providers, on-line publications, and other service providers to sell your personal information to marketers. Service providers will take you off their marketing lists if you request it. Here is how:

1. If you receive junk-mail messages, reply by requesting that the marketer not send you any more messages.
2. If you still receive messages from the offending marketer, you can complain to the "postmaster" at the site where the messages are originating. To locate the postmaster address, drop the first part of the marketer's E-mail address and replace it with the word "postmaster." If, for example, E-mail is coming from Buynowxxx.net, that firm's postmaster's address is probably postmasterxxx.net. Even if the address is wrong, you'll usually receive a response specifying the postmaster address. Ask the postmaster to see to it that the offender no longer sends you junk mail.
3. If the above doesn't stop the unwanted mail, there may still be a way to deal with it. Many E-mail software packages allow users to filter out all messages sent from given addresses.

Messages from "Buynowxxx.net" arrive, but special software automatically deletes them before you see them. Look into the filtering capability of your E-mail system. Shareware programs like eFilter from TSW (http://catalog.com/tsw/efilter), an Internet software design firm, allow you to specify keywords that will automatically delete an E-mail message before it reaches your in box.

What to do if staff does not use E-mail? Try what one of my clients did. The client announced, through E-mail, staff could leave early one Friday. Late in the afternoon this manager walked though the building and there were people who hadn't taken advantage of the free time. They were unaware of the offer because they hadn't bothered checking their E-mail. They didn't make the same mistake twice!

E-mail Rules of Thumb

One of our clients, SmithKline Beecham in Philadelphia, was kind enough to share some of their E-mail wisdom with us. I have summarized some of their ideas and added a few of our own. By using these ideas, you will be saving yourself heartache and extra work.

- In composing an E-mail message, ensure the subject matter on the subject line is obvious.
- Briefly state the purpose of the E-mail in the beginning of the message.
- Limit yourself to one topic per E-mail message.
- Send messages or replies to only those who need to know. Do not use the Reply All key!
- Use the copy and paste feature to add to an E-mail message, rather than attachments.
- If revising or adding to an existing E-mail document, put revisions in bold type so they are obvious to the recipient.

- If someone needs to modify or comment on a document, precede the comments with the author's initials.
- When receiving E-mail, look at the header or subject field of the document before reading the text. Decide whether you want or need to read the whole message or simply delegate or delete it, then and there.
- Turn off the alarm feature. You do not need to be constantly interrupted when an E-mail arrives. Instead batch E-mail by scheduling two or three times a day to process all E-mail at one time.
- Be courteous and forgiving to those who are not.

You can use the Return Requested function of E-mail as a form of documentation of important matters.

Multiple E-mail Accounts

One source of E-mail confusion is multiple E-mail accounts. Having to go through each is tiresome and items are susceptible to being overlooked. If you have multiple E-mail accounts (local area network, America Online, CompuServe, etc.), consider finding one E-mail front end that will receive E-mail from all sources. Possible solutions include Microsoft Office 97 or Endura Pro. Once the front end is set up, all E-mail can be received and sent from one location.

EFFICIENT SEARCHING ON THE WORLD WIDE WEB AND THE INTERNET

If you think you have wasted time in the past finding a paper file, you might be surprised how much more time you can waste trying to find information on the Internet. In fact, surfing the Web can be so time-consuming that a recent survey covered in *Wired* magazine reported

50 percent of regular users of the Web said they simply don't surf anymore—they hit the same sites every time they log on!

You may run into difficulty connecting to the Internet through your service provider at certain high-volume times of the day. It's worth asking your provider when the busiest times are so you can avoid connecting at those times.

This section will cover ways to locate information on the Internet and Web as efficiently as possible. At the time of the writing of this book there is hardly a way to find information easily and quickly on the Internet. But there are ways to cut down on the amount of time wasted.

Using and Organizing URLs

The Uniform Resource Locator (URL) is the address of a Web server or of a Web page. In the case of Netscape, the URL appears in the Location field near the top of the Netscape window.

If you want to bring up a Web page, all you need to do is type the URL address in the Location field and (assuming you're connected to the Internet) press Enter, and the Web site will come up.

If you cannot connect to a Web site, check that you are still connected to your service provider by clicking on the Dial Up Networking window.

Typically the URL has three parts. Let's use my company Web site as an example (see Figure 6.9). Its URL is http://www.ibt-pep.com. The letters "http" stand for Hypertext Transfer Protocol and together with "://www" identify what follows as a Web server or page. The second part of the URL (ibt-pep) identifies the Web server, followed by the third part, the name of the computer network to which the Web server belongs. This can be followed by a slash and more letters naming the Web page and the location on the server.

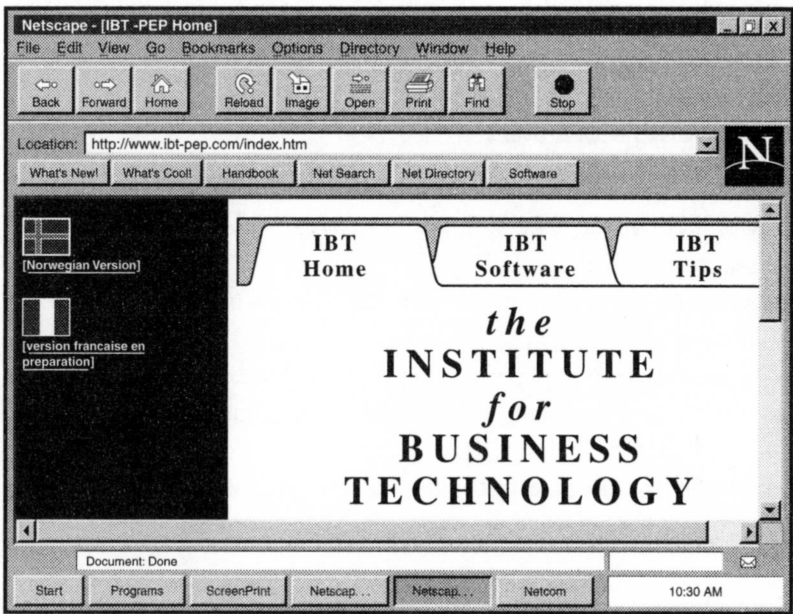

Figure 6.9 Dialogue box from the Netscape Web browser showing the Location field filled in with the Institute for Business Technology's Web site address (URL).

Having a hard time finding a Web site through the URL? Verify the correct spelling of the URL and that periods are really periods and not commas and that slashes are slashes and not back slashes.

Obviously, having the exact site address where you believe the information you need is located is the most efficient way of searching the Internet. So the first rule for efficient searching is: *If you can access the URL for the site that has the information you are searching for, do so.* If you read of interesting sites, note down their URL for future reference. As is the case with an E-mail address, if the URL address is not duplicated exactly when typing it in the Location box, you will not connect to the site. Take care that your address is accurate and that your typing is precise.

Sharing URLs with Others

When you run across an interesting Web site, you may want to share it with a colleague. The simplest way to do so is to send an E-mail message and attach the URL to that message. The recipient can double click on the URL and—if the person is on-line—access the site immediately.

In the case of Netscape, follow the same procedures as for attaching the file, except when you get into the attachment's dialogue box choose Attach Location URL and type in the URL location, making sure to accurately type in the letters and symbols as they appear in the original URL.

Saving Interesting Internet Sites: Bookmarks

Say you are browsing through the Web and run across a cool site you wish to access again in the future. All you need to do is select Bookmarks, which memorizes the Web site address you want to save. Different browser software packages have different names for bookmarks. In Netscape, the saved site shows up under More Bookmarks at the bottom of the bookmark dialogue box. To access the site in the future, simply locate it in your bookmark directory, highlight, and Enter, and the browser brings up the site.

Organizing Bookmarks

You can take this a step further and group similar bookmarks together so it is easy to find a specific site from your bookmark list. Bookmarks can be categorized broadly in the same way as you have categorized and organized your files on the hard drive. You can use the drag and drop feature to move bookmarks and assign them to their proper groupings, so the bookmark organization mirrors the rest of your file organization. (See Figure 6.10.)

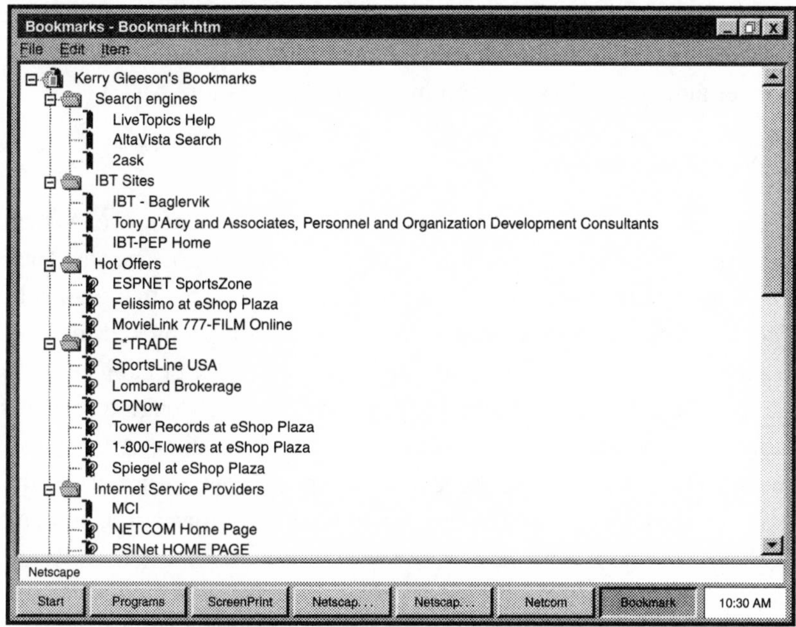

Figure 6.10 Example of bookmark tree.

Search Engines

A search engine is a program that performs a search through most or part of the Internet. Usually it is the best way for you to locate what you want. This is because search engines cover the larger parts of the WWW. This is also the drawback. Estimates say that the World Wide Web now consists of approximately 65 million Web documents. Much of this is of no use whatsoever. But still the search engines spend time sifting through them. And quite a lot of them pop up in your list of hits as noise.

None of the search engines today cover the Net completely. None of them searches through all parts of the documents. I would recommend that you use some of the bigger ones. At the moment these are: Hotbot (http://www.hotbot.com), AltaVista (http://www.altavista.digital.com), and Lycos (http://www.lycos.com). There are also some search engines that activate several engines at once. One of them is Metacrawler (http://www.metacrawler.com). Of course, this takes longer and is not always recommendable. (See Figure 6.11.)

Subject Trees

These are also called subject directories, subject catalogs, or virtual libraries. Subject trees are a Net version of the yellow pages. They have tried to take the part of the librarian. That means that on a subject tree you will find only references that have been picked for relevance and/or quality. They sort their information after subjects with several subcategories. Usually, they have a search engine that enables you to perform a search through their own directory. But beware! The Internet is so huge that the people who administer the subject trees are not able to sort through all the information, and end

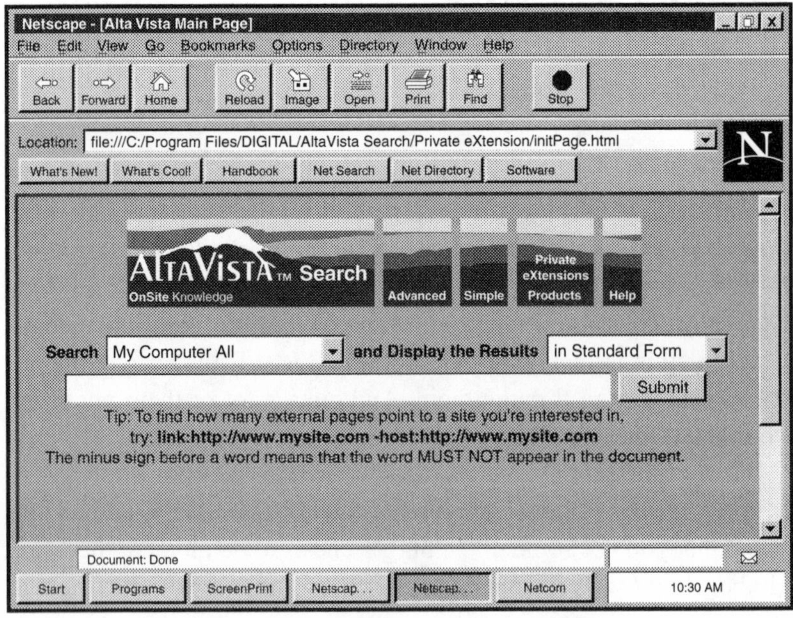

Figure 6.11 AltaVista Search main page.

up just picking the golden nuggets. You will get far more hits by using a search engine. But then you also will get far more irrelevant information. The biggest subject tree is Yahoo! (http://www.yahoo.com). (See Figure 6.12.)

Some subject trees also give you lists of the most popular or newest Web sites. Most subject trees try to cover all kinds of subjects. But there are also some that cover just a narrow field of information. These will often give you more accurate and relevant hits. A list of these specialized trees can be found at http://www.search.com.

Searching Techniques

To understand some of the problems with the Internet, I would like to compare it with a library. The books are put on the shelves according to a system.

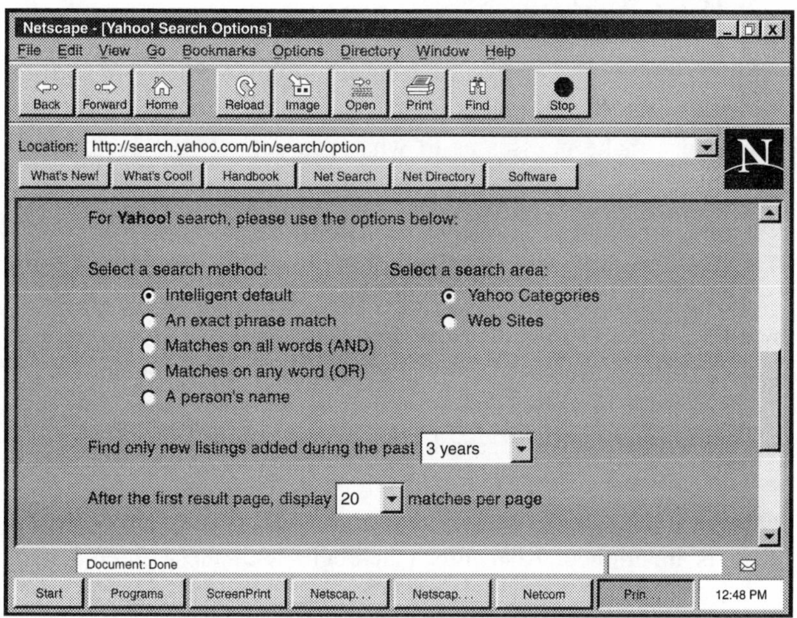

Figure 6.12 Yahoo! search options.

Professional information hunters—librarians—can retrieve them for you. They also act as evaluators concerning what is useful information or what is garbage. The Internet has no limitations when it comes to shelves. There are no librarians who regulate what is available. In fact everyone with a PC and a modem can make available whatever they like, be it valuable or pure trash. That means that you have no guarantee that the information you have found on the Internet is correct. You are on your own out there. The only help you can get is from search engines. (More about those later.)

When you search the Net you will find that you seldom hit the jackpot at once. The information you get can be divided into four categories:

1. Relevant—Exactly what you were looking for.
2. Useful—Parts are of great interest.
3. Marginally useful—Only bits and pieces are interesting.
4. Noise—Not useful at all.

So why don't you just get relevant information when you perform a search? The main problem is that you are dealing with machines. They don't make guesses as to what you actually are looking for. They have no possibility to understand that your needs could be met from another angle. To make the machines into more efficient workers you have to develop skills in searching. Otherwise you will get those 146,349 hits. And there is no way that you want to spend the next five weeks visiting all those Web pages.

As an information hunter my goal is to construct a simple word or string of words that gives me high relevance and few hits. Let's say I use the word "Car" in a search. It turns up innumerable hits—a totally useless list. "Cars" is such a common word that not only will I get information about cars, I will get every text on the Net where that word is mentioned. Also, hits that begin with those three letters: C-A-R. That means items like: "Carson," "Carter," "Carburetor," "Carriage," and so on. The machine takes you literally. Let's look at some techniques that will undoubtedly help you.

If I were looking for information about cars in the United States, I could perform a Boolean search. It means that you only want to retrieve the hits where two or more words that you specify occur. The search-string would be "cars *and* USA," thereby excluding all Web pages that only deal with cars or only deal with USA. But still you would probably get a lot of noise. Could you narrow this Boolean search further? If it's the pollution side of cars that interests you, you could add the word "Pollution" to your string.

Your problem could be solved by searching for an exact phrase. Let's go back to our car example. Perhaps I really was interested in "sports car." By using that term I would narrow the number of hits. Perhaps I could be even more specific and search for "1987 Porsche." Then I would have a search-string that should give me very relevant hits.

After you have spent some time searching for information on the WWW, you will find that in most cases you just have to use either a Boolean or an exact-phrase search. But it might not be good enough. In this case, you have to add some filters.

- *Truncation.* If you are familiar with the use of wild cards, this is just the same concept. It means that you might not be sure of how to spell your search word. And of course you don't want to miss anything. Let's say that you are trying to get some documents about the Italian motorcycle "Ducati." But you just can't remember if it is spelled Ducate, Ducato, Ducata, and so on. To be on the safe side you truncate by writing "Ducat**."
- *Or.* This is a way to combine two lists of hits. I might need information about "Ducati" as well as another Italian motorcycle like "Laverda." My search-string will be "Ducati or Laverda." This is the same filter that allows you to combine pseudonyms in the search.
- *Not.* This is a filter that gives you the opportunity to exclude certain hits. I might want to exclude the sports cars from my car search. My search-string will be "Cars not Sports Cars."

- *Time.* Some search engines give you the opportunity to search for documents only from a specific period. You may only be interested in the more recent information, or you may want everything prior to this year—for instance, information about Porsches from 1990 until today.
- *Geographical.* With this filter I can exclude or choose certain geographical areas, for instance cars in Africa.
- *Media.* This gives you the chance to only seek for specific types of media, like sound-files pictures, videos, and so on.
- *Part of document.* If you know that there is an existing Web page with the word "Laverda" in the address, you can search just in the URLs.

Different search engines have different filters and possibilities for constructing search-strings. You have to make yourself acquainted with each of them and evaluate their possibilities and limitations.

To search an index, enter the terms you are looking for in the Search text box. *Fast Company*, probably the most useful magazine I have ever subscribed to, had a recent article written by John R. Quain, on how to search smarter. Some of the advice he gave included:

- Use quotation marks. Most engines interpret quotes as "search only for sites with all words exactly as typed."
- Use an exclusion. Your first search attempt may yield dozens of similar sites that have nothing to do with what you're looking for. Exclude these listings by adding a minus sign followed by the key words you don't want to see. On one such search engine, for example, "Java" yields 20,000 sites; "Java -coffee" slices the list to 26 sites, all on the programming language.
- In most search engines you can use the search option buttons to indicate whether you want to search the entire search service's index or only that portion of the index that relates to a particular

directory category. Search engines have options that provide additional ways to narrow down your search. In the case of Yahoo!, if you click on Hyperlink (next to the Search command button), the browser displays additional methods for searching (e.g., by name, matching a word, etc.). (See Figure 6.12.) It is possible to have Yahoo! look only at the Web or look only at user groups or look only at unlisted E-mail addresses. The "Find Matches That Contain" option buttons let you specify what happens when you search using two terms, for example, "football" and "Super Bowl." If you only want to find Web pages that use *both* terms, mark the "All Keys" option button. The "Consider Keys to Be" option button lets you specify whether your search terms need to be whole words or pieces of words.

- Each search engine has its own ways of narrowing down your search. In the case of AltaVista, if you want to search for something that's a phrase all you have to do is enclose the phrase in quotation marks.
- If you're trying to search for information about an organization or individual, you might try using WhoWhere Search Service at http://www.whowhere.com.
- Some engines allow for "near" searches. If, for example, you are searching for Paris hotels you may find tens of thousands such sites with the words "Paris and hotels." If you use the "near" word, such as "Paris near hotels," most browsers will bring up sites where the words "Paris" and "hotels" are no more than 10 words apart (the maximum number of words apart will vary among browsers), narrowing down the number of sites that will pop up on the browser.

Specialized Databases

These can be divided into free services or pay services. You can compare the specialized databases with huge encyclopedias that cover one area. The difference between these and the subject trees is that the trees only give you a hyperlink reference. In most cases, however, the databases will give you both the reference to where an article, report, or thesis has been published, and the text itself. Among the thousands of databases available, I will note a few:

- *Uncover.* Gives you the chance to search through articles and reports printed in 17,000 Anglo-American magazines and journals. It's free to search their database. But if you want some of the actual texts you have to pay (http://uncweb.carl.org).
- *Pathfinder.* Here you can find articles and Web pages from 90 magazines, film studios, and other media companies within the Time Warner group (http://www.pathfinder.com).
- *CNN Interactive.* Gives the user a database of local American news or the whole spectrum of international events. Some of the news clips also have links to video and sound (http://www.pathfinder.com).

LET OTHERS SEARCH FOR YOU

I have found that no matter how efficient you might be, finding information on the Web or Internet can still be very time-consuming. One way to avoid this difficulty is to have someone else do the searching for you. A number of different companies, such as PointCast, AirMedia Live Internet Broadcast Network, Microsoft, Netscape Constellation, Wayfarer, and BackWeb, offer services designed to locate and forward information from the Internet.

These services work like an electronic clipping service: They deliver the information you want and have predefined directly to your computer or pager. In most cases, the information is downloaded on to your computer when you are on-line but not actively searching the Web. You can receive information in one of two ways: **push** or **pull**. With "push," the content provider notifies you when new information is ready for you. With "pull," you ask the provider to send any new information. The World Wide Web is no longer only a place to search for information; it can also come to you.

All content providers tend to work more or less on the same principle: You set the preferences for the information you want. But each provider offers a different approach to delivering and alerting you to information.

In the case of **Netscape Constellation**, its In-Box Direct brings the HTML pages you predefine to your Netscape E-mail in box. You can choose from a variety of well-known offerings including magazines like *Salon* and newspapers such as the *New York Times* as well as others.

BackWeb pops up on your screen as a screen saver, delivering small summaries of information. It uses small graphical icons that periodically move across the bottom of your screen. If the message is something you want to look at, you double-click and BackWeb brings up the Web page with more information. BackWeb also uses audio messages.

With **PointCast**, you can also subscribe to selected information (from a large number of sources). PointCast delivers it in an organized way. It provides a window for viewing headlines or summaries and a larger window for viewing the full text. In the screen saver mode, PointCast animates your screen saver with headlines. Clicking on one launches it and lets you view the details.

Many of these services are in their infancy and will improve as time goes by. Of course, they also cost money. But if you need timely information from the Internet and don't have time to do the search yourself, these services come in quite handy.

SAVING AND ORGANIZING WEB DOCUMENTS

Most browser software allows you to organize and save Internet documents (Figure 6.13). Use the basic principles outlined for organizing your desk and your computer's hard drive (see Chapters 2 and 3).

There will no doubt be times when you wish to save information from the browser window. It is possible to save both the graphics and the text from the Web. With many browser software programs, it's also possible to save only text or only graphics from a Web site.

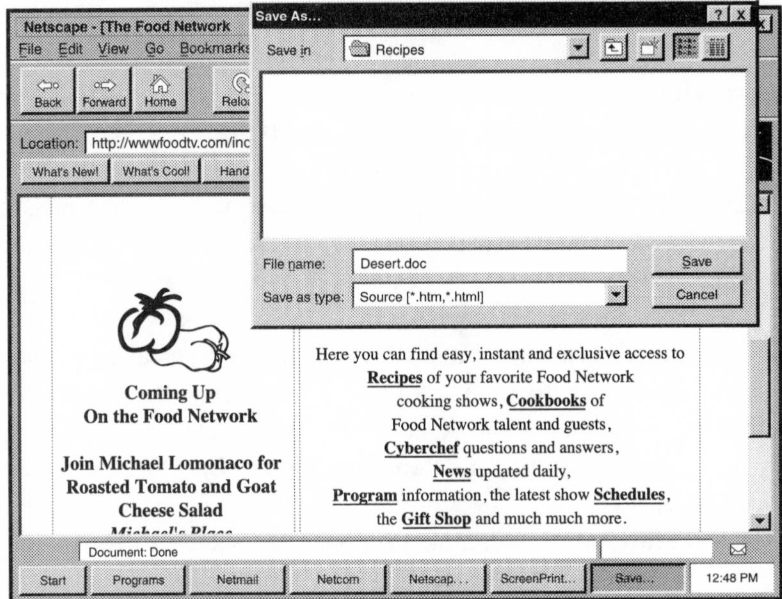

Figure 6.13 Saving an Internet document.

The key to retrieving information you've saved from the Internet is to identify where the document belongs on your hard drive and name the file consistently with the names you used on your hard drive and data files.

SUMMARY

The Internet is a wonderful source of information. It can also be a tremendous waste of time. You can use the World Wide Web more productively by understanding how browser software and the search engines work. By organizing your browser to emulate the rest of the organization of your computer and your paper-based systems, any files you end up keeping are going to be easier for you to retrieve. It is up to you to control the flow of information from the Internet and not let it control you.

FOLLOW-UP ACTION POINTS

✔ **If not already connected to the Internet, do so.**

a. Get the proper hardware (computer and modem).

b. Choose an Internet service provider.

c. Choose browser software.

d. With the help of your service provider, get connected!

✔ **Get familiar with your browser software.** Ask yourself, how can I customize my browser to best suit my work and style?

✔ **Organize your browser E-mail, bookmark, and address folder systems.** Mirror the rest of your desk and computer organization.

✔ **Get familiar with the different search services available to you.**

✔ **Determine if one or more of the many "push" services on the market can provide you with the important information you seek** and at the same time eliminate the need for time-consuming surfing of the Web.

Good hunting!

PREVIEW

Need more than just personal organization? Groupware software is designed to help teams of people organize and share information for a project or common goal.

In this chapter you will learn:

☞ The concepts and terms used in groupware.

☞ The different groupware software that is available today.

☞ The functionality groupware can provide.

☞ How people are using groupware.

☞ How to successfully introduce groupware to a team or to your organization.

7
Groupware

Up until now we have talked exclusively about your personal productivity, work habits, and organization. But if you are one of the many people working in today's organizations that are flatter, with less hierarchy and more cross-department teams, you may be saying, "It's great if I am organized. But so much of my work is dependent on others being organized, and I have no authority over them." Groupware may be the team productivity tool you are looking for.

What is groupware? We define it as computer software explicitly designed to support the collective work of teams and enhance:

- Communications with colleagues and other organizations.
- Collaboration of teams.
- Coordination of business processes and tasks.
- The organization and sharing of information.

Today many companies are moving to flatter organizational structures in order to be more competitive and bring products to market faster. (See Figure 7.1.)

Groupware is a tool that supports changing organizations. If you are working in an organization that is moving toward open sharing of data and information with vendors and/or customers, groupware could be the answer to many organizational and communication issues.

The first groupware software was Lotus Notes. It hit the market in late 1989. Groupware received slow acceptance but is now gaining momentum and is a hot new area for software companies. At this writing there are quite a few groupware products available besides Lotus Notes. They include: Netscape's Communicator, Microsoft Exchange, Novell's GroupWise, Oracle's InterOffice, OpenMind from Attachmate, SuperOffice, and ICL's Teamware.

There are many different functions that we look for in groupware products:

- Electronic mail or messaging.
- Group calendaring and scheduling.
- Group document managing.

Figure 7.1 Adhocracy structure.

- Group conferencing.
- Meeting support.
- Group decision support.
- Information sharing.
- Workflow management.
- Remote access.
- Tools to develop specific applications in these areas.

Not all the products we mentioned have all these functions. But then, you may not need them all.

Before we go any further, it may be helpful to review some definitions of a few terms that are particular to groupware.

Discussion database: a groupware application that facilitates computer conferencing. It allows groups of people to electronically discuss a topic of common interest. There are several major advantages to a discussion database. First, people can participate at their convenience from anywhere at any time. Second, a record of the discussion is kept for reference. Third, only one copy of the information exists (unlike multiple E-mail messages) so it is easier to update and takes up less space on the network.

Replication: a groupware process that updates databases that are located on multiple servers or computers. During replication, database copies are compared for differences. New data is added to all copies of the database and obsolete information is removed. This feature allows users to work remotely without being directly connected to the network server. They periodically replicate to update the server and client databases so that they are identical.

Structured: a method for storing and using data. Transaction systems like accounting or mailing labels are structured where specific information is typed into designated spaces in electronic forms. These forms are used to populate databases. The structured data is then categorized so that it can be searched and stored using predefined criteria (e.g., a vendor name or ZIP code).

Unstructured: a method for storing and using data. Data is entered free-form and does not have specific structural requirements. A word processing document is unstructured data. Searching this data is more difficult than searching structured data. Groupware's strong point is storing unstructured data. Text information can be searched for specific phrases and words. Full text search also allows a user to perform complex searching to find occurrences of words or phrases across many documents.

Threaded discussions: an electronic conference. The discussion subjects are called items. After reading an item the user is prompted for a response which is appended to the original item. Items and responses are searchable. Users are informed of new items or responses that have been added since they last read the discussion.

New responses and items are presented in the order they are added, topic by topic. This is to simulate an ongoing conversation with other participants.

View: a list of documents in a database that is usually sorted or categorized to make finding documents easier. A database can have any number of views: by author, date, subject, and so forth.

In discussion databases, limit the discussion to three levels: a main topic, a response, and a response to the response.

MORE POWERFUL THAN E-MAIL

Groupware is more powerful than E-mail. (See Figure 7.2.) Groupware communicates on a many-to-many basis. For example, a project team has regular meetings on Fridays. Every Monday, the project manager posts the following Friday's meeting agenda in the team's discussion database. Each team member is expected to read the agenda and come to the meeting prepared. If they have any suggestions or want to recommend changes, they add their individual comments to the agenda document. There is only one copy of each document in the database and on the network, yet everyone on the team has access to all the information in one easy-to-get-at electronic location.

E-mail communicates on a one-to-one or one-to-many basis. You send a message to one person and add copies for any others you think might be interested. People respond to you and send copies to anyone else they think might be interested. Pretty soon there are a lot of copies of the same information clogging up your network. And heaven help you if you want to update or correct the information you originally sent. How would you track down all the subsequent transfers? Many groupware applications are "mail enabled," which means you can notify individuals by E-mail that you have added information to a shared database, without actually sending them the information.

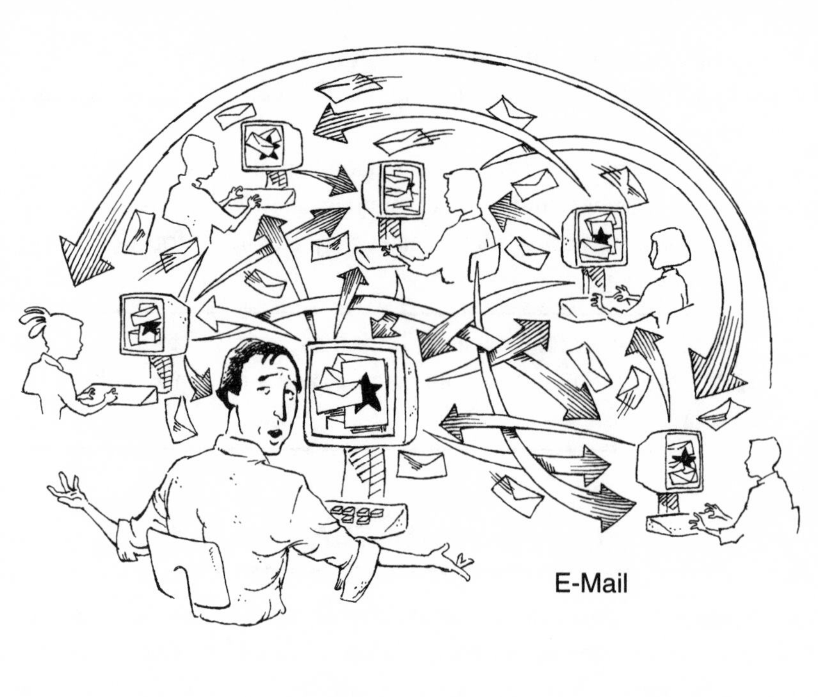

Figure 7.2 Groupware versus E-mail.

WORK ANYTIME, ANYWHERE

Groupware allows people to work anytime, anywhere, even if they are not connected electronically at the time they are working. To give you a feeling for the power of groupware, here's what some people are doing with it:

An association with over a thousand members and a small staff keeps records of all their members, correspondence, and contacts with them. Each staff member is organized with a diary that includes a calendar and to-do list, company-oriented lists of members, lists to track activities, and document templates for correspondence with members. The master database sits on their local area network (LAN). Any contact, whether phone call, meeting, fax, or correspondence with a member, is noted in the database. All staff members carry laptop computers with them when traveling on business. Their information is transferred regularly to the corporate server on their LAN. With groupware, it's possible for anyone in the office or outside of it to know the current status of every project. It's possible to look at others' calendars to schedule a meeting, even if the individual is out of the office. Their document templates give their correspondence and proposals a consistent, professional look.

A pharmaceutical firm uses groupware to manage their internal meetings effectively. Discussion databases are used for all the preliminary agenda creation and for providing background information on the issues. Group calendaring allows efficient scheduling of people and resources like special video equipment necessary for a meeting. Discussion databases are used for follow-up. Actual meetings are confined to decision making and delegation. All those tiresome status meetings are a thing of the past.

A consulting firm wanted to open its first overseas office. It would be staffed with nationals who had no previous relationship with the firm. In order to bring the new employees up to speed, an electronic "library" of all the firm's reports was created. Full text searching across all the reports can be used to create a list of reports that mention a given subject. The names of the team members who worked on the project, the name of the client, and other necessary information is also provided. The actual report can be recreated and edited for reuse. A discussion database allows the users to "talk" electronically with any or all of the team members and learn about other information that may not have been in the report. The fact that the two offices are six hours apart geographically has little bearing on their ability to share information. They have eliminated the need to reinvent the wheel. Arthur Andersen calls this "distributed competency." Experts can reside anywhere. Their knowledge and insight are available anywhere and at any time in the organization.

A consulting team uses groupware to manage a project that involves the client, a software vendor, and an application developer. The client company hosts the network. The other three companies are guests on their network for that application alone. The security of the groupware product they use allows the client to ensure that the other three companies have no further access to information on the client's network. The tasks for the project, any meeting notes, action items, responsibilities, and time tables are all entered into the database. Priorities for tasks are established. There are electronic discussions of ongoing issues and problems among all the members of this multicompany team. When a meeting is necessary, the agenda is established and everyone is notified. Only those team members with agenda items need to attend. The others can review the meeting notes in the database after the conclusion of the meeting.

An international bank has a 12-member department that is spread over three countries and seven time zones. They work in small groups to create documents, which are then distributed within the bank. In order to reduce the amount of time spent traveling, they use groupware to collaborate on their document authoring. The project leader or person designated as author composes the first version of a document using Word, Excel, or another software product. The document is attached to the groupware database, and all the team members are notified of its existence and the deadline for comments by E-mail. Team members comment on the original document in a discussion database, creating a history of everyone's thoughts. The author makes edits, the team comments. This cycle is repeated until the team is satisfied with the document. It is then submitted to the department head for approval. The department head can also use the discussion database to comment on any desired changes. If necessary, the cycle is repeated. When the department head feels the document is final, the head approves it with a distribution list and notifies the author, who then distributes the document using the corporate E-mail system. The team has all documents (in-process and final), a history of all the revisions and reasons for revisions to the document, and a distribution record for their documents in one place. (See Figure 7.3.)

Figure 7.3 Groupware conferencing.

IS GROUPWARE FOR YOU?

As you can probably imagine, with powerful software like this, there are a lot of issues to be considered before deciding to automate with groupware.

Ask yourself this question first: Will your organization really profit from groupware? It should be obvious to you that groupware solutions are successful only when the group is *willing to share information.* We cannot stress this too much. It is a critical success factor. Let's face it: Everyone knows knowledge is power. Anyone not willing to share power is not willing to share knowledge. Such people will sink any attempt to make groupware applications effective. Jim Manzi, the former CEO of Lotus (the company that developed Notes) says that other software applications are centered around information processing. Groupware is centered around "information sharing." If you are one of the lucky people who work in an organization that believes that group knowledge is more powerful than individual knowledge and that the success of a group is success for every member of the group, then you are ready for groupware and can skip to the next section, on implementing groupware.

However, if you work in an organization like many of those we know with "cowboy cultures," many people either are not willing to share information or do not make it a priority. Changing this cultural roadblock may be your biggest hurdle in introducing groupware. You have the age-old options of the carrot or the stick approach. Sometimes a combination of the two is the most effective. One manufacturing company we know of created a groupware application to distribute individual expertise throughout the organization. The chairman was committed to making this initiative a success. He mandated his people use the system. However, even with this kind of pressure, usage was slow in coming, too slow for the chairman. At year-end, the chairman invited the top 150 users of the system to a special meeting. Part of their reward was a new higher-speed laptop and a very expensive leather carrying case for it. These quickly became a status symbol in the company. Usage expanded rapidly after that. The system is now an integral part of the company's infrastructure and is considered a source of competitive advantage.

It certainly helps if a senior person is committed to groupware and uses it. However, the best way to encourage usage is to ensure that the application itself offers some reward. Is it designed to capture information that users really need? Does it cut back on unwanted travel? Does it allow them to work at home several days a week? You have to determine before you begin what will make your people jump at the opportunity to use their groupware application. Make sure that benefit is built in and touted to the group. Include the people who will be using groupware in discussions about design. Let them see a prototype, and get their feedback.

In addition, there should be tangible rewards for users. If this is an important way to increase productivity, decrease time to market, or improve customer relations—things that can be measured at the bottom line—then participants are adding to the bottom line. Put your money where your mouth is. Make participation part of the performance review process in your group and reward those who participate.

IMPLEMENTING GROUPWARE

If you see the power of groupware and feel you have many application areas that are crying out for a groupware implementation, how do you decide where to start? Go with the adage, "you eat an elephant one bite at a time."

First of all, pick the low-hanging fruit. Choose an opportunity where the most improvement or the greatest return can be realized. Choose an opportunity that can be completed quickly, where an application can be rolled out in 90 days or less. This way you will capture enthusiasm and be able to run with it. Remember that groupware in most cases requires a major change in the way a person works. You need enthusiasm for it to be successful. Leave the killer applications until after groupware has become a success and an accepted way of doing business for your organization.

If you have a "cowboy culture" you won't be successful unless you address the cultural issues as well as the technical ones. Thirty percent of the groupware users we have polled have mentioned cultural issues as the major stumbling block to groupware's acceptance in their companies. One successful groupware adopter has the following recommendations:

- *Make it fun.* Have a contest or lottery with prizes for the person who sends the one thousandth message or guesses how many messages get posted a week.
- *Take away the stress.* Let people work at home so they can make mistakes in private or take as much time as they need to "play around" and learn.
- *Appoint leaders.* Whether they are subject specialists or users of the system who can be system specialists, have users available who will hold hands and answer questions.
- *Get everyone on-line.* Senior people particularly should be involved. The message from them is: If it is important to me, it should be important to you.

Another critical success factor in bringing in groupware technology is training and support. This is a new way of work. If you expect people to change lifelong work habits in a short period of time, it is only fair to tell them how.

We feel the most effective training for an extensive change in work styles is to actually simulate the new work, using a case study approach:

- Hands-on, with no more than two people per computer, you give them an actual work problem and walk through the "new solution."
- Give them a similar exercise and let them solve it themselves. The trainer is there to field questions and help discover the answers with the users. But the user does all the work; the

trainer is there as a coach only. The trainer never solves the problem or even touches anyone's computer.

We find that the simulation training method leaves users with real knowledge of the application and how it relates to their own work processes. To neglect training is to scuttle your project before it starts.

Make training fun as well as relevant. Have awards and prizes. Rather than certificates, give participants something noticeable to set them apart.

GROUPWARE IS NO PANACEA

As you read our stories lauding groupware and telling you how powerful it is, you are probably asking yourself, "If groupware is so great, how come more people aren't using it? What's the holdup?" We'll be perfectly honest. Groupware is no panacea!

Here are some suggestions for avoiding pitfalls:

1. Use the most powerful hardware possible. There is no such thing as hardware that is too powerful. If the software manufacturer says 8 MB of memory are minimum, plan on at least 16. Twenty-four or 32 would be even better. Systems that run too slowly frustrate even the most gung-ho user.
2. Create a catalog—a database of databases. Many companies that have let groupware proliferate on a local department level have so many databases of information that no one knows where any specific information is. They are back in the same soup they were trying to crawl out of when they started with groupware. One major accounting firm, an early adopter of groupware, finally asked their librarian to create a database of databases. This same remedy has been employed in other firms as well.

If there is just too much data for you to keep up with, or too many different places to check to see if there is new data, try this solution. Use a groupware function called "mail enabling." As information is generated in a database, the individuals for whom it may be critical are automatically notified by E-mail to check the database. If your applications are not mail enabled, then treat them like your E-mail (maybe even batch them with your E-mail), schedule reading your databases several times a day, and check them on schedule. Don't do it ad hoc—it will either be too frequent or not frequent enough.

3. Get users involved in the decision making process and the development process. Let them own it right from the beginning. Listen to their input. After all, these are the people who will work with the system. Get people to buy in from the beginning.

4. Test, test, test. Make sure everything works as planned before you roll it out to users. We know of one incident where new users inadvertently were given the ability to delete the document links in E-mail. One neat and tidy user deleted the document in his E-mail and, as a result, it was deleted from the database. Clearly the training was insufficient for the technical responsibilities granted. But the outcome was more disastrous than just one lost document. The system lost credibility as all the skeptics said, "It's not reliable! It loses documents!"

5. Train, train, train. In our poll of groupware users, another 30 percent said that lack of or insufficient training was the reason for groupware's slow penetration in their companies. If you have gone to all the work of having an application developed for you or even if you are offering a groupware application "out of the box," offer training. How important can this functionality be to you if you are not willing to train people to use it? Training should be followed by coaching and support.

6. Make sure you are functional as well as portable. Road warriors may need more than a laptop to stay connected. Are cellular modems necessary for truly remote users? Do they travel out of their home countries? They may need different power setups and modems. Fill reclosable bags with the equipment necessary for each country and label it UK, Germany, USA. Get users in the habit of requesting their special equipment packs in advance when they make plane reservations.
7. Make it easy. You are asking for a lot of change. Why make it any harder than it has to be?

Try before you buy! Consider renting an application. For example, you can set up a project management application with several companies or teams in multiple locations in a day and test it before a complete rollout through your company. Check out Involv by Changepoint at www.involv.net.

SUMMARY

Groupware can be a wonderful organizational tool for those who need to share information with others. However, a successful groupware implementation requires commitment, thought, planning, and yes—organization!

FOLLOW-UP ACTION POINTS

If you determine that you, your team, or your organization are candidates for groupware, then:

✔ **Find out if any groupware products are already in use in your organization and what they are.** Share resources whenever possible, list the functions that you and your group need—E-mail, calendaring, shared document writing—and evaluate groupware software against that list.

- ✔ If you decide to implement groupware, **choose a small project with flexible participants** as a starting point. Use it as your pilot. Train all the participants.

- ✔ Based on the results of the pilot, **plan how you will roll out groupware to the rest of your group or organization.**

The material in this chapter was provided by Carter Crawford and Carol Gorelick of Solutions for Information & Management Services, Inc.

Epilogue

I have had the opportunity to work with many hundreds of professionals in scores of countries over the past 15 years. I have been left with many impressions, but a few stand out.

- Despite cultural and geographical differences, most businesspeople, whether employer or employee, believe they have too much to do and too little time to do it in.
- We are creatures of habit. Habits tend to act as barriers when attempting to improve ourselves.
- Most people lack a purpose in life.
- Most work stress is self-created.
- Most people produce but a fraction of what they are capable of producing. They do not recognize that dramatic iprovement in personal productivity is completely under their own influence and control.
- Executives will say they are committed to the development and personal growth of their employees. But when push comes to shove, most simply expect their people to be able to do their job. If they do not or cannot, the vast majority of executives prefer to find some who can, then go through the effort of developing the new people.

- People will say they are committed to continuous improvement, but few spend any time at it.

Here we are, faced with the most rapid technological changes in history, and we stubbornly hold onto work behaviors designed for a much different period.

This observation is not pessimistic. Quite the opposite: I see a wonderful opportunity to gain control over one's work and life! The payback for even modest effort can be outstanding.

No matter where I travel for my business, I find that executives feel compelled to embrace technology. For they are told that if they don't, they will be left behind, and will be of less value to their employers. Technological advances may provide both the tools and motivation we need to become productive powerhouses. What an opportunity!

I say grab it. Use these forces of change to propel you along. Begin by embracing those tools that will provide you with the most immediate results—better control over your time and money.

This book is a working person's guide to getting more done with modern work tools. These tools will continue to evolve and improve. The question you have to keep asking yourself is, will you continue to evolve and improve as well?

Glossary

Active cell The cell in a spreadsheet where you enter data. It has a dark border or appears shaded. Also called the current cell.

Application *A computer program that lets you use the computer to perform a specific task, such as writing a letter, balancing your budget, or playing Rebel Assault. The other type of software, operating system software, works in the background to help the application deal with the disk drives, printers, memory, and any other devices.

Artificial intelligence (AI) *Attempt to make computers and programs mimic human thought processes so they can do more of our work for us. The problem is that computers are logical and humans are . . . well, uuuh . . . , HUMAN!

ASCII file *(*As-key* file) Abbreviation for American Standard Code for Information Interchange file. A code developed to standardize text on all computers. An ASCII text file contains plain-vanilla text—no fancy stuff like margin settings or bold type. ASCII files can be shared by most applications and most computers.

AUTOEXEC.BAT *A DOS file that runs a series of commands whenever you start your computer. Its name explains its function: It AUTOmatically EXECutes a BATch of commands.

Background operation An operation that is occurring in the computer system but is not visible to the user.

Backup 1. A copy of data usually maintained on a diskette, tape, or other medium for recovery if the original is lost or destroyed. 2. An alternative plan. (In order for a plan to qualify as a backup, there must be a primary plan. Plan It Now!)

Base/conventional memory The section of random-access memory between 0 and 640 KB (640,000 bytes) in which most MS-DOS applications or programs run.

Baud rate The speed at which information can be transferred through a COM (serial) port. Expressed in bits per second (BPS), it is the basis on which computer modems are rated for speed.

Binary file A file containing information that is in machine-readable form. A binary file can be read only by an application or program.

Bit The smallest unit of information recognized by the computer. A bit always has a value of either 1 or 0. Eight bits equal one byte (or character). (*See* byte, megabyte, gigabyte)

Bookmark (Internet)—Automatically marks favorite Web sites for easier return and access.

Boolean search Refers to Boolean logic as it applies to data search. From a type of algebra (named after George Boole) which uses binary logic, i.e., the logical operations OR, AND, NOT, etc. It is a search that allows the user to specify the exact information required.

Boot/reboot 1. To start (or restart) your computer loading the operating system (DOS, OS/2, Windows 95, etc.) 2. Abrupt removal from a task, position, or employment, often requiring one to restart.

Browser Software program that enables you to search through the information provided by a specific type of server.

Buffer A section of memory in a printer that stores information that is selected to print.

Bulletin board (Internet)—Computerized versions of the old cork bulletin boards that serve a similar function: You can leave messages or advertise what you want to buy or sell.

Byte The amount of space required to store a single character, such as a typed letter (or space) from your keyboard. A byte consists of 8 bits.

Cache memory An area of very fast memory dedicated to storing data retrieved from main memory. Data is temporarily stored in the cache memory in anticipation of future use by the microprocessor. A cache memory will significantly increase the speed of most program operations.

Cascade A way of arranging open windows on the desktop so they overlap with the title bar of each window remaining visible.

C drive In an IBM PC or compatible computer, this is usually the first hard disk drive in the computer. You can partition this hard disk so it acts as several drives. The computer assigns a letter to each partition (C, D, E, F, and so on), but you still actually have only one hard disk drive.

CD-ROM (Acronym for Compact Disc—Read Only Memory) Type of optical disk similar to an audio CD, but with a different track format capable of holding 600 to 650 MB of digital data (including text, graphics, and hi-fi stereo sound). "Read only" indicates that data can be read but not altered.

CD-ROM drive Compact Disc—Read Only Memory drive. A device that reads information stored on compact discs.

Cell A single box in a spreadsheet or table.

Cell reference Defines the location of each cell in a spreadsheet or table. It consists of a column letter followed by a row number (example: A2). Also called a cell address.

Central processing unit (CPU) The computer's brain. It processes instructions, performs calculations, and manages the flow of information through a computer system. Also called a microprocessor. The CPU in a personal computer contains a single microprocessor chip that's slightly bigger than a poker chip and costs much more.

Character *Any letter, number, or symbol (#!@$%*96ag>~&) that you can type.

Characters per inch (CPI) *The number of letters that occupy one inch in a line of text. Common measurements are 10 CPI and 12 CPI. The larger the CPI, the smaller the type.

Chat rooms (Internet)—"Virtual rooms" where users can communicate with each other by typing in real time.

Clipboard A temporary Windows storage location used to transfer data between documents and between applications. The Edit, Copy command, (or Ctrl + C) copies selected items to the clipboard. Edit, Paste, (or Ctrl + V) pastes items from the clipboard to the position of the cursor.

Close command To remove a window or dialogue box, or quit an application. When you Close an application window, you quit the application.

Communications program *An application that lets your computer and modem chat with other computers and modems. Communications programs usually contain the commands that tell your modem how to dial the phone, how fast to talk, and what language to use, so you don't have to worry about such nonsense.

Compatibility The ability of two or more to exist together in a common situation without having a major problem. File compatibility means that files created by one program can be opened and edited by the other. Command compatibility means that commands in all subject programs are similar.

COM port (serial port) A connection on a computer where you plug in the cable for a serial device (serial printers, modem, etc.). Windows supports COM1 through COM4.

Compound document A document that contains information created by using more than one application (program). Example: a word processing document containing a spreadsheet.

Computer A general-purpose machine capable of doing a number of things. Unfortunately it usually requires some input from a human being.

Computer for the road (Not without a designated driver!) *See* Personal digital assistant (PDA) and Notebook (laptop) computer.

CONFIG.SYS A file that includes commands that tell the operating system how to function. When you turn on a PC, it automatically reads the CONFIG.SYS file and executes the commands, which may tell the system how to manage memory, files, and devices.

Continuous incremental improvement (CII) A self-motivated search to maximize one's potential through continual upward goal setting. "If you want to keep getting what you're getting, then keep doing what you're doing."—Les Brown.

Conversion The process of transforming a file into a different format so that it can be opened and used in a program other than the program used to create it. Whenever you create a file (such as a letter) in a program, that file is saved with codes that tell the program how to process the file. If you wish to open the file using another program, the file must be converted to a format for that program.

COPY command To put a copy of the selected text or item onto the clipboard so that you can transfer it to another location. Most Windows applications have a Copy command on the Edit menu that performs this task. You can also copy entire files from one location to another by using File Manager.

Cryptography The science of encoding messages or electronic files for the purpose of increased electronic information security.

Cursor A flashing line on a computer screen. It indicates where information you enter will appear. Also called an insertion point.

Database A program that helps you manage large collections of information.

Data file A file where data information is stored. The product of word processors, spreadsheet programs, database applications, and so on.

Data processing *Using the computer to type, rearrange, and fetch information. Although all aspects of computing are technically data processing, the term usually refers to information-intensive work with large databases.

Data storage An area where data is kept or stored. A diskette, fixed disk, tape cartridge, and CD-ROM are examples of data storage media.

Default The value that will be used by a computer system or program unless the user alters it to use another value.

Default button In some dialogue boxes, the command button that Windows automatically selects. The default button has a bold border, indicating that it will be chosen if you press the Enter key.

Desktop An on-screen work area that is supposed to resemble your cluttered desktop. The first screen to open, it typically contains a menu, program groups, icons, and so on, all designed to simplify life. Most such programs (MS Windows, OS/2, etc.) provide selected tools (calendar, notepad, phone lists), that will allow the user to open more than one program at a time. Simplicity and organization is the key to desktop efficiency.

Device A single piece of hardware that performs a specific function (printer, disk drive, mouse, etc.).

Device driver The software that provides an interface between an operating system and a device. It is either built into the operating system or must be installed using the CONFIG.SYS file.

Dialogue box A small, on-screen window that lets you select options for a specific task.

Directory Like a folder in a filing cabinet, a directory organizes information stored on a computer.

Directory/subdirectory A way of grouping files. Each directory contains the name and size of each file and the date and time it was created or last changed. It will also contain the names of other subdirectories. A subdirectory is a directory which is subordinate to another directory.

Discussion database A groupware application that facilitates computer conferencing. It allows groups of people to electronically discuss a topic of common interest.

Disc/disk *A disc is a bologna-slice-shaped device used to store data. This spelling is usually used for compact discs. Don't ask why! A disk is a flat, round piece of plastic covered with magnetic particles that stores data, sort of like a cassette tape. Normally, you won't see the round part that's the disk. The disk is permanently enclosed in a square protective sheath.

Disk cache Part of a computer's main memory. It stores data recently used by a computer to improve its performance.

Disk maintenance *The proper care of a disk and disk drive. Although disks are generally reliable for storing information, you must maintain a disk to make sure it remains reliable and efficient. This is especially true for hard disks. To keep the disk in tip-top shape requires a maintenance utility software such as PC Tools or Norton Utilities. (*See* Disk maintenance 2.)

Disk maintenance 2 *Keep your disk organized. Keep related groups of files in separate, logical directories. This way you can find files easily. If you need to delete files, you can delete an entire group.

Disk maintenance 3 *Keep your disk tidy. If you don't use a file, copy it onto a floppy disk and then erase the file from your hard disk. If you don't use a program, back up the program's files and delete them from your hard disk.

Disk maintenance 4 *Back up regularly. Get a backup program and perform weekly and/or daily backups of your hard disk. An up-to-date backup can protect you in case any files are deleted or damaged.

Disk maintenance 5 *Defragment files. (Optimize disk.) Over time, files become fragmented on the disk. Each file is no longer stored in one contiguous area on the disk. The result is that it takes your hard disk longer to read files from the disk, and it makes it more likely that parts of a file will get lost or damaged. Defragmenting rearranges the files on the disk.

Document The product of an application created by the user and stored as a file on a disk (business report, memo, spreadsheet, picture, etc.). *Also see* File.

DOS (MS-DOS) (Microsoft Disk Operating System) A versatile but somewhat difficult operating system for computer systems having disk capability. Usually the system found on IBM-compatible computers. DOS is to Windows 95 as a six-speed stick shift (with no A/C or radio) is to a loaded Mercedes Benz. (*See* Operating system.)

Dot matrix printer The least expensive type of printer. It is ideal for printing on multiple forms, which need an impact pressure to print through multiple copies.

Double-click To rapidly press and release a mouse button twice without moving the mouse. Most mouse software will permit the user to adjust the sensitivity or speed of the double-click required.

Download The process of copying electronic data or files from another computer to your own.

Drag To move an item on the computer screen by holding down the mouse button while moving the mouse pointer to the desired new location. You can move a window to another location by dragging its title bar.

Drag and drop A method of moving files or selected information from one location on the screen, in a document, or in the file directory to another location. The item is captured by pointing to the item, using the computer mouse, pressing and holding down the left mouse button. While holding the button down, move the pointer to the new position and then release the button.

Electronic mail (E-mail) *Mail that requires no postage and usually gets there on time. E-mail is a system in which people can send and receive messages through their computers on a network or using modems. Each person has a designated mailbox that stores messages sent by other users and can then retrieve and read messages from the mailbox.

Expansion card A circuit board that adds a new feature, such as CD-quality sound, to a computer.

Expansion slot A socket inside a computer where you plug in an expansion card.

Extension (file name extension) The period or dot and up to three characters at the end of a filename. An extension usually identifies the kind of information a file contains.

File A collection of information for use by the user, or if in program language, for use by the computer to execute operations.

File compression *To squeeze files so that they take up less space. People commonly compress files to fit more of them on a disk or send more data in less time to another computer. In either case, a file compression program uses a type of shorthand to reduce the size of the files. The files then need to be decompressed in order to make them usable.

File Manager *A program or a feature of a program that allows you to copy, move, and delete files on disk so that you don't have to use DOS. Many file managers also allow you to manage directories on a disk; you can add, delete, or move directories to restructure the directory tree. File Manager lets you do all this and more without leaving Windows. Jump at anything that helps you avoid DOS.

Filter *Any process that allows only specified information to be processed, displayed, or printed. The filter screens out unwanted values, names, or other data.

Floppy disk A removable device that magnetically stores data. Also called a diskette.

Floppy drive A device that stores and retrieves information on floppy disks.

Folder *A group of files and/or folders that is represented onscreen by an icon of a folder. In Microsoft Windows, File Manager displays directories and subdirectories as folders.

Font A set of characters of a particular design. The sizes of the characters are measured in points. (There are approximately 72 points in one inch.)

Gigabyte (GB) An amount of storage equal to exactly 1,073,741,824 bytes. It is commonly thought of as a thousand megabytes or one billion characters. (*See* byte.)

Gopher Internet databases that can be accessed by the Web as well as other gopher clients.

Grammar checker *A program that checks for incomplete sentences, passive voice, awkward phrases, subject-verb disagreement, wordiness, and other grammar problems. In my opinion, grammar checkers are unreliable and cranky. You'll learn more about grammar by reading a Virginia Woolf novel.

Groupware A software tool which allows the open sharing of data and information between individuals, even if they are not connected electronically at the time they are working.

Hacker *Slang term for someone who is particularly skilled and knowledgeable about computers and who loves to play on computers. Often used to describe someone who can perform unusual tricks with a computer, such as breaking into other computers.

Hard drive 1. The primary device that a computer uses to store information. 2. From New York to Los Angeles, nonstop . . . , in a pickup.

Hardware *The physical equipment that makes up a computer, including the disk drive, monitor, keyboard, mouse, modem, printer, cables, and any other cool electronic stuff. These items are distinguished from software: the instructions that tell the computer what to do. Think of it in human terms: Your hardware is your body and mind; your software is your education.

Home page (Internet/intranet)—The first screens that appear when visiting particular Web sites.

HTTP Hyper-Text Transfer Protocol, the format of the World Wide Web. When a browser sees "http" at the beginning of an address, it knows that it is viewing a WWW page.

Hyperlink A colored section of text (usually blue) that, when clicked, will take you to another Web page relating to that subject.

Hypertext *A method of presenting text in which selected text is linked with key terms. It is commonly used in on-line help systems to provide you with a quick way to move through the system. The chosen help system provides information about that topic and one or more highlighted terms that act as links to related topics. Clicking on a link provides information about the related topic.

Icon A small picture on a screen that represents a program, command, file, or other object. You normally select an icon by clicking on it, or activate it by double-clicking.

Image That intangible physical element that determines or influences the mental picture for observers. It has been said that "image is everything."

Index *An alphabetical list of terms and corresponding page numbers that appears at the end of a document. In most word processing programs, you can create an index by marking each occurrence of the term you want to index. Once the terms are marked, all you have to do is generate the index and tell the program where to put it.

Index file *A separate file that keeps track of all the records in a database file. Because a database file can be very large, some database applications use an index to search for and sort the records. This increases the speed at which the program can find and sort records, but it requires that the index file be updated regularly.

Integrated circuit (IC) *"I-see," nicknamed microchip. A tiny chip (usually made from silicon) that holds a full network of electronic circuits working together to process data. It was this technology that made personal computers a reality, because it shrank the size and increased the speed and reliability of computers so remarkably. *See* CPU, a type of IC.

Interface *A connection that allows you to interact with the computer and allows the computer to interact with peripheral (external) devices. The most common interfaces are: hardware—cables, connectors, and ports that link computers to devices (printer, modem, mouse, etc.); software—commands and codes between programs; user—devices used to communicate with the computer (monitor, keyboard, mouse, etc.).

Internet A network of computers from around the world connected to each other. Originally developed by the U.S. Defense Department to link scientific research institutions through their computers, it has grown by millions of computers, and is now used by millions of people to get and share information, access the WWW, send E-Mail, and so forth.

Intranet An internal network built on the same technology that makes up the World Wide Web, specifically HTML (Hyper-Text Markup Language) and HTTP (Hyper-Text Transfer Protocol). Intranets may be run over local area networks and wide area networks as well as public networks.

ISDN Abbreviation for Integrated Services Digital Network. Basically allows a faster transfer of digital data, provided compatible equipment is used.

Keyboard The device connected to most computers that is used to enter information and instructions. It includes standard typewriter keys and specialized keys.

Kilobyte (KB) A standard unit used to measure the size of a file, the storage capacity of a disk, and the amount of computer memory. A kilobyte is 1024 bytes and is equivalent to 1024 characters.

Local area network (LAN) A type of network that connects computers within a small geographic area, such as an office or building.

Location boxes (Internet)—Spaces on the Web browser where you type in the Uniform Resource Locator (URL) address to the site you want to visit.

Megabyte (MB) An amount of storage equal to exactly 1,048,576 bytes. It is commonly thought of as one million characters. (*See* byte.)

Megahertz (MHz)/clock ticks The "clock" regulates the pace of the microprocessor. The clock is rated on the number of ticks or cycles per second. One megahertz means one million clock ticks per second. A higher clock speed means a faster microprocessor.

Memory *What you lose when you have kids. In the computer world, memory is an electronic storage area that forgets everything when you turn off the juice. Whenever you run a program or open a file, the computer reads information from disk and copies it into memory. Why? Because a computer can work with the information much more quickly when it's stored electronically. (*See* RAM.)

Modem A word fabricated out of MOdulator—DEModulator. A computer hardware device that lets computers exchange information through telephone lines.

Modem, fax A device that allows document transmission to and from fax machines and other fax modems.

Monitor A device that displays text and graphics generated by a computer. Looks like a television set without knobs, switches, speakers, or remote control.

Motherboard The main circuit board of a computer. All electrical components plug into the motherboard. (Not to be confused with mother's board, which was sometimes used to punish.)

Mouse 1. A handheld pointing device used to select and move items on a screen. 2. A small, ubiquitous rodent made famous by a Disney character named Mickey.

MS-DOS Shell MS-DOS Shell is a graphical, menu-driven DOS management tool that lets you operate in DOS using few keystrokes or a click of the mouse. It is used not only to organize files and directories, make copies of files, run programs, and so on, but also to provide a visual display of the directories and files on the disk. Can be opened by typing DOSSHELL at the DOS prompt.

Multimedia *The combination of text, sound, video, graphics, and animation for use in presentations, games, and computer-based training.

Nerd/geek *A nerd is similar to a geek but a lot less fun; nerds love to work on computers just for the sake of working on them. They'll do stuff with the computer that is very complicated but extremely boring. Geeks do complicated but interesting stuff. Both nerds and geeks have little interest in human beings or even in being humans.

NerdPerfect A rare day without crashes, clashes, or blowups. A near-perfect techno-day.

Network 1. A group of computers connected together to share information and equipment for mutual benefit. 2. A group of business entities connected together to share information, services, and so on, for mutual benefit. 3. A group of anything connected together to share anything for *any* mutual benefit.

Neural network *An artificial intelligence system that is designed to mimic the activity of a human or semihuman brain. (*See* artificial intelligence.)

News groups (Internet)—Discussion groups devoted to talking about a specific topic.

Notebook computer (laptop computer) *Also called a luggable, a lightweight computer that you can carry around like a briefcase that's stuffed with barbells. It is complete with a display, keyboard, and processing unit and is battery-operated. Generally called notebooks when under 10 pounds.

Operating system (DOS, OS/2, Windows) Software that controls a computer by managing programs, input/output, scheduling, data management, and interaction with mass storage devices.

Path *A map that tells your computer where to look for a file. In DOS, you can organize your files in directories and subdirectories. Whenever you want to open a file, you must tell DOS where the file is located. To do that, you must specify a path. The path tells DOS which drive to start with and which directories to follow to get to the file.

Pentium *Intel's successor to the 486 microprocessor chip, the Pentium processes instructions between two and five times faster than the 486. Another significant advance is that the Pentium offers two memory caches (the 486 has only one). One cache is used for data, and the other is used for program instructions.

PEP Acronym for Personal Efficiency Program. A worldwide structured program designed to help professionals to work more efficiently. For more information, contact:

IBT International, Inc.
P.O. Box 1057
Boca Raton, FL 33429
Phone: (561) 367-0467
Fax: (561) 367-0469

PEP talk Personal Efficiency Program time-saving jargon such as:

1. Do it now! Tomorrow it may be a crisis.
2. A stitch in time saves a trip to the tailor.
3. Never put off for tomorrow . . . anything!
4. Plan to plan!
5. If you don't have time to do it right, you won't have time to do it over.
6. Take time to sharpen your *ax* . . . (or *acts*).

Peripheral A piece of hardware attached to a computer, such as a printer or keyboard.

Personal digital assistant (PDA) A portable, battery-operated computer that fits in the palm of your hand. It may serve as a general-purpose computer but is usually preprogrammed to perform a few specific tasks. The most popular PDA, the Apple Newton MessagePad, includes a notepad, address book, fax, calendar, and to-do list. It has optical character recognition capable of translating handwritten notes into typed text.

Personal information manager (PIM) A term commonly used to describe software designed to help organize personal information. Usually contains a calendar, to-do section, address/contact section, a planner, and some type of reminder alarm. Examples include Lotus Organizer, MS Scheduler Plus, Side Kick, etc.

Port A socket at the back of a computer where you plug in an external device, such as a printer.

Printer *A device that turns the masterpieces you create on your computer into paper documents the whole world can enjoy. There are a number of printers to choose from, each with a cost roughly indicative of the print quality. (*See* dot-matrix, inkjet, laser, and thermal printers.)

Printer, dot-matrix Inexpensive and noisy, and provides a low to medium output quality. These printers create characters and graphic objects as patterns of tiny dots.

Printer, inkjet *(Such as the BubbleJet and DeskJet) Offers a good mix of affordability, quality, speed, and silence. An inkjet prints by spraying ink onto paper. Sound messy? It can be if you print on absorbent paper. However, with the right paper, an inkjet printer can produce a printed page that would stand proud next to a laser-printed page.

Printer, laser *Top-of-the-line printer, making it the highest priced of the bunch. A laser printer works sort of like a copier machine to produce slick, professional-looking printouts quietly and quickly.

Printer, thermal *Inexpensive and provides a low to medium output quality; however, it is quiet. Thermal printers apply high temperatures to expensive heat-sensitive paper, producing print that is hard to read and smells funny. These printers are usually used in fax machines. Few people would go out of their way to buy one.

Procrastination Procrastination is the art of keeping up with yesterday.—Don Marquis.

Even if you're on the right track—you'll get run over if you just sit there.—Arthur Godfrey.

PROcrastinator A person who has taken the art of crass neglect to the highest level and has forfeited amateur status.

QWERTY *Pronounced "kwer-tee," the standard typewriter keyboard layout. The name QWERTY refers to the first six letter keys in the upper left of the keyboard.

Random-access memory (RAM) Memory that is used to run applications and perform other necessary computer tasks. (*See* virtual memory.)

Random-access paralysis (RAP) A mental condition that occurs when demands for one's personal memory exceed one's memory capacity. Often identified by the sound "Duuhhh . . . !" Can be avoided by the use of PEP principles.

Replication A groupware process that updates databases that are located on multiple servers or computers. During replication, database copies are compared for differences. New data is added to all copies of the database and obsolete information is removed.

Root directory The highest directory of a disk created when the disk is formatted. The root on the C drive is indicated by "C:\".

Search engine (Internet)—A database or index that can be queried to help you find information on the World Wide Web. For example, WebCrawler and Yahoo!

Security *There are various ways of securing your system and files, including protecting files with a password, locking your computer with a key, and even using data encryption techniques such as those used by the government.

Shortcut A Windows 95 feature that allows the user to quickly access selected documents or frequently used programs by placing their shortcut icon on the desktop. It does not change the location of the object; it just lets you open the file quickly by double-clicking the icon. (You can also put a shortcut in any folder, in addition to on the desktop.)

Simulation An application (program) that lets you use your computer to pretend you're flying an airplane, driving an Indy car, or really working.

Software *Also known as programs and applications, the human-developed instructions a computer needs in order to operate. Without instructions, the computer has no life of its own and is just a lump of circuitry in a metal case. There are two basic types of software: operating system software and applications. The operating system gets the computer up and running and applications tell it how to perform specific tasks.

Static RAM (SRAM) *A fast, expensive type of random-access memory that is commonly used for caches. Unlike DRAM (dynamic RAM), which needs to be constantly recharged to remember data, SRAM remembers as long as the power is on.

Structured A method for storing and using data where specific information is typed into designated spaces in electronic forms. The structured data is then categorized so that it can be searched and stored using predefined criteria (e.g., a vendor name or ZIP code).

Subdirectory *A subdivision of the current directory. Because large hard disks can store thousands of files, you often need to store related files in separate directories on the disk. Think of your disk as a filing cabinet and think of each directory as a drawer in the cabinet. A subdirectory is equivalent to a group of files in the drawer, for example, all files under the letter K.

Surfing (Internet)—Similar to television channel surfing. Allows you to travel to different Web sites with clicks of the computer mouse.

Swap file A file that sits on a computer's hard disk and helps the computer imagine that it has more memory. The computer shuffles information between the swap file and RAM as needed.

Tape drive A device that lets you copy the information stored on a computer to tape cartridges. Also called a tape backup unit.

TCI/IP (Internet)—A communication protocol by which most Internet activity takes place. Access to this protocol allows multiple simultaneous connection to any services on the Internet.

Threaded discussions An electronic conference. The discussion subjects are called "items." After reading an item, the user is prompted for a "response," which is appended to the original item.

Unstructured A method for storing and using data, which is entered free-form and does not have specific structural requirements, as in a word processing document. Text information can be searched for specific phrases and words.

URL (Internet)—Uniform Resource Locator, an HTTP address used by the World Wide Web to specify a certain site.

Utility programs Programs that help with the periodic maintenance and troubleshooting of your computer systems. Two such programs are PC Tools and Norton Utilities.

View A list of documents in a database that is usually sorted or categorized to make finding documents easier.

Virtual memory A memory management system used by Windows 3.1 that allows Windows to behave as if there were more memory than is actually present in the system. It is equal to the amount of RAM plus the amount of disk space allocated to a designated swap file.

Virtual oversight The result of failing to accomplish a task or acknowledge an event in a timely fashion although it was not actually forgotten. Tends to make one appear much more ineffective and forgetful than one really is. Can be overcome by practicing PEP principles.

Virus A disruptive program created by mischievous computer vandals. For example, it can display annoying messages on a screen or destroy information on a hard drive. Computer viruses can be transmitted from one computer to another through shared programs or diskettes. Not harmful to humans. (No microorganisms are involved.)

Wildcard Special symbol that stands in for other characters and is often used to search for or select a word or file name. Usual wildcards include the question mark (?) for single characters and an asterisk (*) for multiple or a string of characters. An example of the use is **.exe*, which selects all files with an "exe" extension in a directory or folder.

Word wrapping A feature in a word processor that automatically moves words to the next line as you type.

World Wide Web (WWW, W3) *An Internet tool that lets you skip around from article to article till you find what you need. Each article contains one or more key terms that cross-reference other articles. Key terms may be underlined or followed by a bracketed number, such as [1] or [2]. When you select a key term, its article is displayed on-screen.

Zip drive Plug-in-and-go hardware that is like three drives in one—it enables unlimited hard drive expansion; it is portable like a floppy disk, but holds a lot more stuff at hard disk–like speeds; and as a backup system, its 100 MB disk allows you to consolidate up to 70 floppy disks on a single zip disk.

References

Bliss, Edwin. *Getting Things Done: The ABC's of Time Management.* New York: Scribner, 1976.

Covey, Steven R. *The Seven Habits of Highly Effective People.* New York: Simon & Schuster, 1989.

Drucker, Peter. *The Effective Executive.* New York: Harper & Row, 1966.

Gookin, Dan. *PCs for Dummies.* (Foster City: IDG Books, 1996).

Hafner, Katie and Lyon, Matthew. *Where Wizards Stay Up Late: The Origins of the Internet.* New York: Simon & Schuster, 1996.

Hill, Napoleon. *Think and Grow Rich.* New York: Fawcett Crest, 1960.

Hobbs, Charles. *Time Power.* New York: Harper & Row, 1987.

Kraynak, Joe. *The Complete Idiot's Guide to Computer Terms*, 2d ed. Indianapolis: Alpha Books/Macmillan Computer Publishing, 1994.

Mayer, Jeffrey J. *Time Management for Dummies*. (Foster City: IDG Books, 1995).

McCay, James T. *The Management of Time.* (Englewood Cliffs: Prentice Hall, 1959).

Morris, Larry. *E-Mail and Messaging.* (Indianapolis: New Riders Publishing, 1994).

Negroponte, Nicholas. *Being Digital.* New York: Vintage Books, 1995.

Nelson, Stephen L. *The World Wide Web for Busy People*. (Berkeley: Osborne/McGraw-Hill, 1996).

Senge, Peter M. *The Fifth Discipline*. (New York: Doubleday/Currency, 1990).

Seymour, Jim. *Jim Seymour's PC Productivity Bible*. (New York: Brady/Simon & Schuster, 1991).

Winston, Stephanie. *Getting Organized.* New York: Warner Books, 1978.

SELECTED COMPANIES/PRODUCTS MENTIONED

Apple
  Newton MessagePad (hardware—PDA)
AT&T
  Pocket Net Phone
Colorado
  Jumbo External Tape System
Computer Insurance Agency (service)
Coral Graphics
DayTimer Technologies
  DayTimer Online
Digital Equipment Corporation
  Office of the Future/Flexible Office
Intuit
  Quicken (software—financial)
Iomega
  Zip drive (hardware—auxiliary drive/backup)
Lotus Development Corporation
  Agenda (PIM—Obsolete)
  Lotus Organizer (PIM)
  Freelance (graphics software)
Microsoft
  Windows® (operating systems software)
  Schedule+ (PIM)
  PowerPoint (graphics software)

Millennia Software
  Email Reader
Mitsubishi
  Wireless Communicator Mobile Access
Motorola
  PageWriter
  Personal Messenger 100C Wireless Modem Card
National Dispatch Center
  Socket Wireless Messaging Services (SWiMS)
Nokia
  Nokia 9000/PCS 1900 (phone)
Pilot (hardware—PDA)
PK Ware
  PKZIP (utility software)
PSION (hardware—PDA)
Sharp
  Wizard (hardware—PDA)
Skytel
  Skywriter (pager)
Socket Communications, Inc.
  PageSoft Pro
Traveling Software Corporation
  Laplink (utility software)
TravelPro suitcase (hardware protection)
Unwired Planets
  UP.Link
Wildfire (electronic voice messaging system)
Wynd Communications Corporation
  WyndMail
ZAP-it

Index

About the Institute for Business Technology

Should you wish to order a free copy of any of the forms found throughout this book or purchase the PEP Planner software, please write or call:

The Institute for Business Technology International, Inc.
P.O. Box 1057
Boca Raton, FL 33429
Telephone (1) 561-367-0467
Fax (1) 561-367-0469
E-mail address: kgleeson@ix.netcom.com

If you would like more information about the Personal Efficiency Program or if you would like to speak with a PEP representative, contact one of the following Institute for Business Technology offices:

The Institute for Business Technology, Australia and New Zealand
P.O. Box 6199
Shopping World
North Sydney NSW 2060, Australia
Telephone (61) 2-9553269
Fax (61) 2-9555480

The Institute for Business Technology, Belgium
Lange Nieuwstraat 58 Box 14
2000 Antwerp, Belgium
Telephone (32) 32263610

The Institute for Business Technology, Benelux
P.O. Box 688
1180 AR Amstelveen
Westelijk Halfrond 487
1183 JD Amstelveen, Netherlands
Telephone (31) 20-6473752
Fax (31) 30-6477633

The Institute for Business Technology, Canada
96 Donegani, Suite 604
Pointe Claire
Quebec H9R 2V6, Canada
Telephone (1) 514-6319207
Fax (1) 514-6316284
E-mail address: pep@ibtcda.ca

The Institute for Business Technology, Denmark
Lyngso Allé 3
DK 2970
Horsholm 400, Denmark
Telephone (45) 45-762512
Fax (45) 45-762520

The Institute for Business Technology, Germany
Wilhelmstr. 43
58332 Schwelm, Germany
Telephone (49) 233693900
Fax (49) 2336939030
E-mail address: IBT-PEP@compuserve.com

The Institute for Business Technology, Italy
Via M Gonzaga 7
20123 Milano, Italy
Telephone (39) 2 86398211
Fax (39) 2 72001339

The Institute for Business Technology, Mexico
1651 Scooter Lane
Fallbrook, CA 92028
Telephone (1) 760-731-1400
Fax (1) 760-731-1414
E-mail address: ibtwest@earthlink.net

The Institute for Business Technology, Norway
Gamle Brevik VE-1
N-1555 Son, Norway
Telephone (47) 64-959225
Fax (47) 64-959162
E-mail address: johholst@telepost.no

The Institute for Business Technology, Sweden
Kraketorpsgatan 20
431 53 Molndal, Sweden
Telephone (46) 31-7061950
Fax (46) 31-877990

The Institute for Business Technology, Switzerland/France
Case Postale 339
1224 Chene-Bougeries, Geneva, Switzerland
Telephone (41) 22 8691100
Fax (41) 22 8691101
E-mail address: ibt@iprolink.ch

The Institute for Business Technology, United Kingdom
P.O. Box 95
Dorking
Surrey RH4 3FR, England
Telephone (44) 1306887944
Fax (44) 1306884161
E-mail address: help@ibt-uk.demon.co.uk

The Institute for Business Technology, United States
P.O. Box 1057
Boca Raton, FL 33429
Telephone (1) 561-367-0467
Fax (1) 561-367-0469
E-mail address: kgleeson@ix.netcom.com

The following consulting firms are licensed to sell and deliver IBT services:

HIPRO Consulting
Samdo officetel #1106
12-1 YOIDO
Young deung po
Seoul, Korea
Telephone (82) 27611080
Fax (82) 27611082

Trends International Marketing Co. Ltd.
3F-1, No 79
Hsin Tai Wu Road
Sec 1, Hsichih
Taipei County, Taiwan
Telephone (886) 26984889
Fax (886) 26984896

Should you wish information on groupware consulting services, please write or call:

Solutions for Information & Management Services, Inc.
5 East 16th Street
Suite 650
New York, NY 10003
Telephone (212) 675-4747
Fax (212) 675-5479